Backyard Livestock

Backyard Livestock

HOW TO GROW MEAT FOR YOUR FAMILY

STEVEN THOMAS

Drawings by Mark Howell

The Countryman Press

WOODSTOCK · VERMONT

Fifth Printing

Library of Congress Cataloging in Publication Data

Thomas, Steven.
Backyard livestock.

Bibliography: p.
Includes index.
1. Stock and stock-breeding—Handbooks, manuals, etc.
I. Title.
SF65.2.T48 636.08'83 76-43244
ISBN 0-914378-21-X
ISBN 0-914378-11-2 pbk.

Printed in the United States of America

Contents

Dedication

For Meme, who fed the animals while I wrote about them and without whom none of this would be fun.
and
For Peter and Jane Jennison, for all their help and encouragement over the years and for their friendship.

Discontent is the want of self-reliance.
—*Emerson*

Preface

The land is a mother that never dies.
—*Maori*

The reports are ominous. Drought continues in the sub-Sahara; the Soviet Union has suffered yet another dismal growing year. Even now farmers in the American Midwest are feverishly harrowing their fields to prevent precious top-soil from being blown away by strong winds. There is already talk of another dustbowl like the ones that occurred most recently in the thirties and fifties. For those who are confident that technology will again bail us out, little comfort is given by the fact that climatologists see more such disastrous farming weather in our future than we've had for years—even centuries. They describe the exceptional growing weather we have enjoyed since the thirties as just that—exceptional. Such consistently ideal weather for growing has not occurred in the last thousand or more years, and it is not expected to return for a like number. It is not very comforting to realize that we have based our lofty estimates and stocks for the future (as well as seeds, soil science and planning) on such atypical weather. How will we grow enough food for the future?

If the world's population were declining—even leveling off—there would still be cause for great concern. However, the population is increasing at a rate of 2% per year. As I write this (March, 1976) there are close to 4 billion people in the world. Every *month* that passes since I wrote this the population will increase *7 million*—almost a quarter of a million people per day! At the present rate of growth the population will double every thirty-five years. The population will increase 4 billion to 8 billion by 2011, and in the next thirty-five years it will increase 8 billion to a total of 16 billion in 2046!

Population pressures have resulted in the farming of marginal land and the overgrazing of livestock with the result that in drought years these lands may become desert. A United Nations study indicates that as much as 6.7 percent of the earth's surface may be man-made desert.

Increasing shortages and constantly escalating meat prices are inevitable because of drought, population pressures and higher feed prices. My answer to this problem, obviously, is to grow your own meat. To those of you who doubt your ability to do this, don't forget that it has only been in the last 50 years or so that much of the human race has given up growing its own food. Is it hard work? Is it time-consuming or just plain drudgery? Hardly. On the scale we're talking about—growing meat for one's own family—the investment in time and effort is miniscule compared to the rewards. To raise a pig, a couple of lambs and some laying hens the daily time requirement would be less than half an hour. Your animals needn't cut down on your mobility. If you want to go away have a neighbor take care of your stock; you can reciprocate when he goes away. We have baby sitters—why not animal sitters? And you cannot put a price tag on the pleasure you derive by doing it yourself, not to mention the superiority of your own products over their store-bought counterparts. It was enlightening for me, having grown up in New Jersey, erstwhile Garden State, to discover that a lamb chop is a naturally-occurring item found along the backbone of a lamb, not a commodity wrapped in plastic, nestled next to Extra-Lean Ground Chuck in Aisle Six.

Frances Moore Lappé in her book *Diet For A Small Planet* calls meat animals a protein factory in reverse, because a commercially-raised steer, for example, consumes over 21 pounds of *humanly-edible* plant proteins to produce just *one* pound of meat for human consumption. Therefore, critics say that the real answer to the world's food problems is to give up meat. Fine. Millions of people are vegetarians by custom or preference, but most Americans would sooner give up Apple Pie (and even Mother) before their meat. The answer is to grow meat more efficiently, using wasted food, pasture, or otherwise unused foodstuffs—and that, in part, is the aim of this book. Even Ms. Lappé admits that meat animals *need not* be wasters of food, that they can indeed be "protein factories." Ruminants (sheep, goats, cattle), because of bacteriological action in their stomachs, can convert low-quality feed into high-protein meat. They can graze secondary or unused pastures, vacant fields that lie fallow in wait for industrial parks, and even suburban lawns. The other animals and poultry we will discuss can make similar use of unused or wasted food or feedstuffs you can economically grow yourself to produce meat and other products for your family.

Let's face it: The United States is a wasteful nation. We throw away bottles, we throw away cans; using some pre-infantile logic we extol the virtues of a lighter that, still in working order, we can throw away without having to refill (Hooray! Another breakthrough!). Most of all, we waste FOOD—food that is humanly-edible or could at least be fed to animals which in turn would produce meat. Perhaps you've never really thought about it, but consider how much food your family wastes: dumping a half-finished bowl of soup, throwing away that piece of meat that won't quite make a meal, throwing away glasses of milk, pieces of cake, etc. The best-fed animals in the world are America's garbage disposals. Think of the *tons* of food thrown away daily by schoolchildren, by supermarkets, by restaurants . . . your family. Just think.

The *Vermont Natural Resources Forum* reports that Dr. William L. Rothje, a Harvard archaeologist, has been studying the behavior of Americans by examining their garbage. In Tuscon, Arizona, he found that the average person throws away 42 pounds of *edible* food per year—that's one-tenth of all food brought home! If 15 people saved their wasted food, they could raise a pig and grow almost 200 pounds of pork FREE. Take restaurants: it is their business not to waste, but they still throw away tons of food that is edible by animals (and in some cases humans). We save food scraps from one small restaurant, in one very small town in one of the smallest states—Vermont—and feed it to our pigs. Last year on those scraps alone we raised *three* 250 pound pigs that furnished us with over 500 pounds of delicious hams, bacon, sausage, and pork chops. FREE. On what would have normally been thrown away to rot.

Many of us have grown up in a time of plenty, free from hunger, and think nothing of waste. Others began their lives in a time of need and reason, "I worrried about it then, but now I can finally afford not to." In both cases the result is WASTE. It is time for that to stop, and that is another aim of this book.

Right now, the "Backyard" concept is limited to those rural, semi-rural, or "exurban" areas where animals are allowed, or at least not dis-allowed. If enough people want it, restrictive suburban zoning can be liberalized. Animals are not carriers of disease; they are not filthy—only their owners are. The much-maligned pig is one of the cleanest of all livestock and, if given the chance, might have taken better care of the Hudson River than we did. My sister and brother-in-law, living in suburban New Jersey, have about 50 pigeons, a flock of laying hens, and a horse. The neighbors complained at first, probably only because it was illegal, although they were never bothered. When they were offered the horse periodically as a free lawnmower, they were delighted. The fresh eggs were likewise welcomed. The fact is, if properly cared for and in sane numbers, animals can only be an asset. "But what will we do with their manure?" If you lived in the country the question would be "Where can we get more"—for the garden, or to sell for extra cash. In time when gardens will be more of a necessity than a hobby, manure will be coveted. I think it's a new wave; it will be happening more and more. If you wake up some morning to the sound of a lamb bleating across the road or a chicken clucking as it scratches for food outside your window, be happy your neighbor cares. That's one more mouth off the food chain—better for you, better for everyone. In former lean times politicians promised "a chicken in every pot;" those campaigning in the near future will do well to promise a lamb on every lawn, chickens in every lot.

Many people have contributed their time and knowledge to help in the production of this book. The author would like to express his gratitude especially to the following: my editors, Jane Jennison and Marguerite Sheffield, for laboring over at times illegible typescript to make this work into a cohesive book; Henry and Cornelia Swayze, for their own information on sheep; Patricia Bryson, for her information on pigs; Bill Sumner, Windsor County (Vermont) Agricultural Extension Agent, for his advice on many aspects of this book; Drs. Jim and Steve Roberts of the Woodstock Veterinary Clinic, for many hard-to-find facts and their ongoing help with our animals; and all the people who took the time to send me feeding routines that they have found to be helpful in raising their livestock.

S.T.

Introduction

Through want of enterprise and faith men are where they are, buying and selling, and spending their lives like serfs.—*Henry David Thoreau*

I think the best advice to give anyone beginning to grow their own meat is to *start slowly.* Raising animals is a most enjoyable and satisfying experience, but it can be a nightmare if you get in over your head. We started a bit too fast: we had a couple of horses and some chickens; then we got a couple of pigs, some sheep, and before we knew it the Vermont winter was almost upon us and we had no suitable shelter for our now pregnant animals. We did, somehow, manage to get our animals settled for the winter (our pig two days before she farrowed), but we vowed not to get any more animals until we had pens constructed, and we fully realized how much additional time each animal would take. The following table will give you a rough idea of how long each animal (in the case of poultry, a dozen birds; sheep, 6 to 12 head) will take of your time each day, summer and winter:

TABLE 1: Approximate Time Per Day To Care For Livestock

Animal	*Time/Day (Minutes)* Summer	Winter
Poultry	5	7
Sheep	5*	10
Milk Goat	15-30	15-30
Pig	5	7
Rabbit	5	7
Veal Calf	7	10

* If you raise grass lambs that are contained by fencing you will have practically no time investment (except to check their general well-being) other than to bring water. If you grain them in the summer figure an additional few minutes.

The difference in time required between summer and winter reflects the need to carry and thaw water and to clean pens (since animals will spend more time indoors in the winter months). The time you'll need to care for your stock will depend on how far you have to carry water, how well organized your pens are, etc. Don't let the amount of time scare you if you're figuring on keeping a number of different animals. If you have 3 pigs, a flock of chickens and a dozen sheep, it probably won't take a full 7 minutes per pig, plus 10 for the sheep and 7 for the chickens in the winter for a total of 38 minutes per day. There is a lot of overlap: a big pail can be filled up to supply water for all the stock; pigs are fed the same ration, near each other or in the same pens. Above all, with experience you will organize your barnyard, or yard, so your time will be used most efficiently.

The kinds of animals I have included in the book were chosen for ease of care, for relatively small space requirements, desirability of products and, most importantly, their ability to forage for their own food or to make use of low-protein food, otherwise wasted foodstuffs, or the products of other animals. The conventional beef animal, for example, wasn't included because of its poor feed conversion ratio.

The most important feature of this book, in my mind, is the section *Supplementing Commercial Feed* found in each chapter under the *Feed* section. Here I have attempted to give you some independence from the feed companies by pointing out other feeding routines you can implement yourself. Let's be pessimistic and assume there'll be grave problems feeding the world in coming years . . . or, if you will, let's just be practical: feeding your animals without buying commercial feed is a cheap way to eat! This is just a beginning: I have collected some methods, either from my own or other people's experience, of alternative feeding. I should warn you, though: I have not personally tried *all* the supplemental feeding routines, so use some caution when implementing them. Best of all, use them as a starting point in developing your own routines. I make this appeal: if you know of, or develop any effective and economical feeding routines for any of the animals in this book, please send a complete description to:

"Feeds"
c/o The Countryman Press
Taftsville, Vermont 05073

and you will, at least, receive acknowledgement in any future editions of this book. I believe this is an exciting prospect and hope we can gather and share such information.

I have tried to organize this book, with its chapter divisions and subdivisions, to make it easy to refer to any aspect of livestock care. Each chapter is divided into nine major sections, listed below with a brief description of each:

(1) **Breeds:** I have not tried to acquaint you with all major breeds within a species; however, the most common and useful for a backyard operation are discussed, descriptions are given and in some cases sketches are provided.

(2) **Purchase:** The how and when are discussed as well as the whys of purchasing a certain animal. Features to look for, and avoid, are offered for each animal. This also includes information on how to care for the young animal. In the case of poultry, hatching as well as brooding facilities and methods can be found. To get a first hand idea of what a good specimen of a given species looks like, go to 4-H clubs, livestock shows or county fairs and inspect prize-winners.

(3) **Housing:** Simple shelters for summer raising as well as more elaborate winter quarters are described. Also included are details of fencing and other methods of confining your stock.

(4) **Equipment:** Feed dishes, water systems, and other day-to-day needs are discussed and simple plans shown to enable you to build your own. More specialized equipment for breeding and health care are covered in their appropriate sections.

(5) **Feed:** This section is divided into (a) Conventional Feed and (b) Supplementing Commercial Feed. The first subdivision covers those feeding practices using commercially-purchased feed, hay or the like. When buying your feed be aware of the going prices. Feed stores in one area often charge sizable differences in prices for the identical or comparable feed. Feed is usually cheaper in summer months. Whenever the price takes a sharp drop, buy in quantity if you can afford it and can store it properly. Save your feed bags; most feed companies will pay you a quarter per bag if you return them, although they rarely tell you this. The proper watering of stock cannot be stressed enough. Water should be available, thawed, *at all times*.

Supplementing Commercial Feed has been mentioned above. In addition, topics in this section include mixing your own feeds, free or very cheap sources of feed that can be substituted for commercial feed, and general hints on how to stretch your feed dollar and make your meat cheaper. If you focus on collecting supplemental feed—and make a real commitment—you will be amply compensated by cheaper meat costs.

(6) **Management:** This section has three subdivisions: (a) Routines; (b) Handling and (c) Predators. It also contains miscellaneous information that does not comfortably fit in any other section.

"Routines" deals with the general management of your animals. Suggestions are made as to when to raise particular animals, in what numbers, and how to establish the most efficient and money-saving regimen of raising an animal for your family.

"Handling" involves just that—the correct (and easiest) ways to catch, transport or carry particular animals. Believe me, this can be a godsend.

"Predators" points out the most common animals that will prey on your stock and how to prevent or eradicate them. As a rule, a good barn cat will do a good job protecting your investment in animals and feed.

Keeping accurate records is an important facet of any livestock raising operation. Those "bargain" eggs you think you're getting may not turn out to be such a big deal when you actually calculate how much feed is going into the production of each dozen eggs. It is easy to fool yourself if you don't keep records. You might remember only four bags of feed you bought for your pig when you actually used eight—keep track. I like to keep a separate book on each animal (or flock, herd, whatever) to see how they pay off in the long run. Depending upon your aims—profit, break-even or just good eating—you can make an accurate evaluation of how each animal is producing and decide on the most efficient routines for your family. These records should also contain other pertinent information, such as breeding dates and outcomes, butchering weights, overall cost per pound and the like. These records are also fun to look back on over the years to see how efficiently you've learned to run your operation. It's also nostalgic to look back at how cheap feed used to be. (They are also necessary if you are filing income taxes as a farmer.)

You will discover in time that the concept of backyard livestock makes for better stock. In raising animals, smaller numbers are best. Diseases are reserved, for the most part, for animals that are overcrowded and lack proper sanitation. You shouldn't have any such problem with a backyard flock. In larger operations an owner simply doesn't have the time to check each animal daily and separate those animals that aren't competing well for food. In a small operation you will have few, if any, low grade animals because you can offer individual attention. If one of three pigs is being bullied you can, without much bother, make another pen for it or simply feed it at some distance from the other two. A large-scale owner must accept such poor competitors or take a financial loss and cull him. I can see it with our sheep: when we had a half dozen they all got their share. As our flock grew, and separate feedings and other "coddling" became impossible, "poor competitors" became apparent. I had to take measures to ensure they all got their share or resign myself to a few unthrifty sheep. In short, our animals are superior because I can watch each one carefully and spot any "bullying," signs of disease, or other problems early and correct it before there is any permanent damage. In a larger operation, this is far more difficult and time consuming.

In managing your animals, try to spend some time with them every day. The first sign of disease is often a drop in feed consumption, and if you're attentive you'll spot this right off. By standing and watching our animals as they run in the barnyard or graze, I can spot even the slightest abnormalities: foot problems, lameness, cuts, or other abnormal behavior. Spend some time with them and, above all, *be nice to them.* They are furnishing you with a most precious product. Pet them, talk to them, scratch them—never be mean to them—and they will pay you back.

(7) **Breeding:** This section has details on the breeding practices of all stock (except turkeys and veal calves). It includes information on when and how to breed, gestation, feeding during pregnancy, equipment for breeding and birth as well as weaning and castrating.

(8) **Health:** The key to guaranteeing the health of your animals is PREVENTION. It is repeated so often throughout the book that you may get sick of it, but it does bear repeating. If you furnish your stock with proper amounts of

food and water, clean quarters and fresh air and sunshine you will, unless there's a local epidemic of an infectious disease, have a disease-free operation. This section lists the most common afflictions such as parasitic infestations (worms), their prevention and treatment. A disease table for each animal including major afflictions, their causes, symptoms, prevention and treatment is located in the *Appendix* for easy reference. These charts should aid your quick diagnosis and treatment of major diseases. If you are really serious about raising livestock, you would do well to purchase one of the veterinary manuals listed in the *Appendix*.

(9) **Butchering:** Complete butchering information is given for all poultry and rabbits. The procedures for the remaining animals are too detailed to be given adequate coverage in this volume. If you watch them being butchered once or twice, and use one of the books on butchering listed in the *Appendix*, you should be able to master the procedure yourself.

You might be able to sell or barter some of your surplus meat to further reduce your costs, but check state laws first. Many states require meat for sale to be butchered only in state registered slaughterhouses. You do, however, have the additional option of selling the animal alive and having the purchaser arrange his own butchering.

The last chapter, "Grow Your Own . . ." is designed to be used in conjunction with the "Supplementing Commercial Feed" section and gives you an introduction to growing your own grain crops, pasture, and making hay and silages. There are also references in the *Appendix* you can use to obtain additional, detailed information on growing your own feedstuffs.

I have attempted, within sane limits, to make this a one-book reference. I hope it will suit both the person who wants merely to raise a few laying hens and a lamb and a pig to those who wish to get more deeply involved: raising animals on a fairly large scale, arranging their own breeding and doing their own castrating, butchering and the like. For those subjects I cannot cover because of personal ignorance or limited space, I have included an *Appendix* listing books for further reading. There is also a Glossary, and appendices on administering an injection, constructing and operating an incubator, and on how to tan animal hides. The *Appendix* tables on common diseases are accompanied by tables furnishing information on manure, normal body temperatures, age of puberty, gestation periods, average productive life and related matters.

We are lucky to have livestock and poultry that consistently grow fast, live long, are disease-free, have large and healthy litters, etc. If anyone were to ask me if there is any sure way of raising superior livestock, I'd have to say yes. Simply: *do it by the book* (I'm not being immodest—you needn't use this book—do it by *any* book). Do what you're supposed to do. Furnish your stock with plenty of fresh water, a proper and clean pen and the correct amount of feed and that's it. The stock I see having problems usually aren't getting water or their owners are taking other shortcuts. Such people suppose that, for example, by cheating on feed a bit the pig will grow anyway and be cheaper. Not so. If it says to feed your nursing pig ten pounds of feed a day—*do it.* Sure, she'll probably make it on seven pounds; maybe won't even lose any pigs, but she'll be run down, have a shorter productive life. Then see if you get a reputation for having good piglets!

Lastly, I don't profess to be an expert. Not by a long shot. Since I have finished this book I have learned countless things I should have included. Many people know more about animals than I do and have raised them for more years, but I guess they didn't want to write about it. There are no "experts," you'll learn every day no matter how long you've raised animals. In that sense it's presumptuous to think that this book is complete—refer to other books in the *Appendix,* pump other people for information, keep your eyes and ears open, and you'll be surprised at how fast you'll learn. It may have taken you eight years to learn your multiplication tables, but you'll learn about farrowing the day your sow is due.

CHAPTER ONE

Poultry

I. CHICKENS

When my wife and I made our decision to try to become self-sufficient, our first thought was "chickens!" There was a poultry farm not far away, so we tossed a large box in the back of the station wagon and were off. Within a few hours our yard was graced by six pullets and an uncommonly beautiful rooster. As we took the box from the car we noticed an added bonus: on the bottom of the box was our first egg!

Chickens are the choice of many people when they want to begin raising their own livestock and it is a good one. As opposed to larger stock they need little land, generally no fencing and, depending on your ingenuity, a modest cash outlay for housing. You can purchase a small laying flock cheaply, and it will begin to produce almost immediately. Chickens are among the easiest of creatures to care for, and as your knowledge increases you can easily begin hatching and brooding your chicks. Chickens are easy to butcher and dress. You also have the perfect specimen to acquaint yourself with the basics of livestock raising.

BREEDS

There are almost 200 different breeds of chickens listed in the *American Standard of Perfection*. If you peruse a poultry catalog (see *Appendix*), you will be amazed at the many unusual and downright bizarre breeds of chickens, but for our use we'll classify chickens into three groups (1) "egg" birds; (2) "meat" birds; and (3) "dual-purpose" birds. The so-called egg birds are small birds and are excellent egg producers, but because of their size are rarely raised for table use. Meat birds are large, broad-breasted birds that convert most of their feed to meat and are as a result poor egg producers. Dual-purpose birds incorporate the qualities of the egg and meat birds into one and enable the homesteader to make an enjoyable meal out of any egg producer that is foolish enough to stop laying.

TABLE 1: The Three Classes of Chickens and Common Breeds Within Each

Egg Birds	*Meat Birds*	*Dual-Purpose*
Leghorn	Cornish	Plymouth Rock
Minorca	Orpington	Rhode Island Red
Ancona	Australorp	New Hampshire
Blue Andalusian	Brahma	Wyandotte
Bantam	Cochin	
	Langshan	

White Leghorn

Rhode Island Red

Orpington

Barred Plymouth Rock

Of course, your choice of breed will depend on the needs of your family. Once you taste rabbit you may simplify selection by getting a few layers and relying on your rabbits for meat, but that's another consideration altogether. In dual-purpose birds I have found Plymouth Rocks superior to New Hampshires and Rhode Island Reds in meat quality and about equal in egg production. The White Plymouth Rocks dress out more attractively than the Barred Plymouths with no dark feather shanks left in the skin but you can often sell the neck feathers and skin* of the Barred Rock roosters to sports stores for use in fly tying.

By experimenting and trying other kinds of birds (as in other aspects of raising your food), you will find out which are best suited to your needs and taste. Most commercial egg and meat

* To prepare it take the skin off the neck with feathers intact and salt to preserve it.

birds are now hybrids. Our first birds were a Bantam—Rhode Island Red cross. They incorporate the docility and egg-laying ability of the Bantam with the somewhat larger eggs and better meat quality of the Rhode Island Red. For the ambitious poultry raiser, cross-breeding opens up an interesting opportunity for experimentation. I remember a story in the news a few years ago about an Australian teenager who, by crossing at random different breeds of chickens, came up with a bird that matured to 26 pounds and laid one-pound eggs! Panic-stricken poultry companies offered a million dollars for the bird so they could *destroy* it and with it any competition for their rubbery chicken and mass-produced eggs. Since I have heard nothing else about it (and have yet to see 25-pound chickens in the supermarket), I can only assume the boy accepted their offer and now lives a life of leisure munching buckets full of Kentucky Fried Chicken.

PURCHASE

To decide on the number of chickens you'll need, take into account your egg and meat consumption and determine whether the birds are intended for home use or whether you plan to defray their feed costs by selling surplus eggs. As an aid in helping you decide the number of birds you need for your particular situation, figure a good layer in her first year of production will give you 200 to 220 eggs. The first year of production is the best year, and each succeeding laying year a hen produces a smaller number of eggs. That decrease depends on health, breed, age and most of all the care you give her. Generally, you can depend on almost one egg per day per bird until she goes into molt (see *Management* section), when production usually ceases altogether. As far as meat is concerned, again depending on breed, you can expect a bird to dress out at between two to five pounds at the age of five or six months. If you plan to market eggs in your neighborhood, you can buy as many as you have space for and can handle financially, but don't undertake more than you think you can market. However, if you are just starting out, it is best to fill your family needs and worry about selling eggs and/or meat at a later date.

Once you have decided on your breed and quantity you are ready to make your purchase. You will have the option of buying (1) mature birds (i.e., birds of production age or older);

(2) day-old chicks, or (3) fertile eggs for hatching. If this is your first venture, or you're unsure about the time and effort you wish to expend, I would definitely recommend the purchase of birds that are already in production. This frees you from the problem of hatching or brooding chicks (plus the wait of four to six months for your first eggs!), and the discouragement of losing them at this young age due to disease or beginner's ignorance.

Mature Birds

If you take this option you are ready to gather farm fresh eggs the next day (or as in our case, on the ride home). Often, depending on your source, it may be cheaper to buy mature birds rather than feeding them for the five months it takes for them to begin producing. You might purchase them from a nearby poultry farm, a livestock auction, or from an overstocked neighbor. Commercial poultry farmers don't usually keep their birds past their first year and you can often get a real bargain on birds in their second year of production. They are not worthwhile to the profit-conscious poultryman, but their slightly lower production will be more than offset by their lower price. The disadvantages of buying mature birds are that you will probably be limited in your choice of breed, and if you're not careful you might buy poor laying or nonlaying birds. Table 2 outlines characteristics to look for in choosing a good laying hen. This can be used both in purchasing your birds and in culling birds from your existing flock.

TABLE 2: **Characteristics of Good and Poor Layers**

Characteristic	*Good Layer*	*Poor (or Non-) Layer*
General health and appearance	Alert, well proportioned, active	Deformed, listless, weak
Eyes	Bright and alert	Dull, listless
Comb and wattles	Thick, smooth and bright red	Pale, shrunken, dry, scaly
Vent	Large, moist, oval; white or pinkish	Small, dry, round and yellow
Pubic bones	Flexible, wide apart: at least three fingers in width	Hard, two fingers or less apart
Abdomen	Soft and pliable, thin-skinned	Firm, thick-skinned.

Pigmentation	Pigment bleached from vent, earlobe, beak and shanks (legs)	Yellow pigment in these areas
Plumage	Bright, smooth, clean	Dull, matted, dirty

When you buy your birds it would be wise to buy a colorful young rooster. There's nothing quite like the crow of a rooster in the morning, unless of course he's directly outside your bedroom window. Also, he is half of the means of self-generating your next flock of chickens.

Day-Old Chicks

Buying just-hatched chicks enables you to have a wider choice of breeds, to watch them grow and have that inexplicable joy of finding that first walnut-sized egg five months later. Chicks are usually sold in three classes: 95 percent pullets, as-hatched (about fifty-fifty), and cockerels. You will pay the most for pullets, and the least for cockerels, with as-hatched falling in-between.

Nothing against Sears, but while their *Farm Catalog* is a good source of information on some breeds of chickens, you would do better dealing directly with a hatchery (which is precisely what Sears does, and adds to the price for their profit). You might try looking for a hatchery in your telephone book, or you can surely locate one in the back of a farm-oriented publication (see *Appendix*). Another excellent source of day-old chicks are feed stores. Often in the spring as promotion for their feed they will give away 25 cockerel chicks of a meat-type with the purchase of a bag of feed.

Chicks should be ordered only from flocks that are certified to be pullorum-free. Pullorum-infested chicks (see *Health* section) are often unthrifty and mortality can be quite high. Chicks that are shipped are especially vulnerable so deal with certified hatcheries. Unless the hatchery is local the chicks will be sent through the mail. Don't shudder at this—for some reason the never-reliable mail service excels when delivering day-old chicks. (Perhaps the answer to poor mail delivery is to enclose a chick with every piece of correspondence.) After hatching, a chick has enough nourishment for two days, which is plenty of time for delivery. You will be notified by the hatchery a day or two before your order is shipped and, miraculously, the next day your post office will be filled with the sounds of life. Your next step, *immediately,* is to put them in a brooder.

Brooding is the process from birth to about six weeks of age whereby the chicks are kept confined (but not crowded) in gradually decreasing heat, with plenty of food and water and, above all, strict sanitation. It is the artificial process similar to the natural brooding you see as a mother hen stands in the yard with a clutch of chicks huddled beneath her.

The basic requirements in any brooder setup are:

Adequate space. One-half square foot per chick up to a month of age and one square foot thereafter until taken from the brooder.

Correct heat. 90°-95° the first week with a lowering of five degrees each week until 55° or the outdoor temperature is reached, whichever comes first.

Adequate food and water.

Adequate ventilation without drafts.

Must be predator-free.

I have seen much more elaborate recommendations, including those suggesting infrared heat lamps or gas-fired heating elements, but as long as the basic requirements are met there should be no problem. For small groups of chicks we have simply used a cardboard box with a light suspended in it and placed it in our mudroom. For larger numbers I built a frame in the corner of an unused room and placed an old screen door on top. (The upper half of the door was screened and the lower half wooden.) I simply placed a 60-watt light bulb in the lower end and allowed the chicks to regulate the heat themselves by moving closer to or farther away from the light. Their food and water was placed in the screened end. The screening on top let in light and also allowed for good ventilation without being drafty. Another type of brooder setup for a large number of chicks is the hover type. A hover with an infrared lamp can either be suspended from the ceiling or set up off the floor with blocks so that it allows the chicks free access to the heat. Heat requirements are the same as listed above and the chicks can be allowed to use it free choice. Each chick should be allowed seven inches of hover space.

Litter with large-sized pieces, such as wood chips, peanut hulls or the like are preferred as they help maintain even floor temperature, make for good absorption and allow larger pieces of manure to fall through. In warmer weather, it is essential to check the temperature in the brooder so the heat doesn't get too intense. During a particularly hot and humid day it will be fine to turn the light off until the temperature cools down in the evening.

Too much light and heat and overcrowding in a brooder can result in cannibalism. For some reason the chicks will start pecking at each other, usually around the eyes and the vent, and when blood is drawn all will turn on that particular chick and peck it to pieces, then turn to the next unlucky one. A simple preventive measure for cannibalism is to have the chicks debeaked before you buy them. This involves cutting ¼ inch off the top beak soon after hatching. This service is almost always available from the hatcheries and costs about three cents per chick. If you can't get your chicks debeaked the elimination of the above-mentioned causes (i.e., too much light, heat and overcrowding) should prevent any problems. Perhaps one breed is more susceptible than another, but we've never had this happen and I'm sure we've violated most of the preventive measures, except overcrowding. Some people suggest using a light bulb no brighter than 15 watts, but we've always used 40 watts or higher. I would submit that the most critical factors are a combination of overcrowding and lack of food. If, despite all precautions, you still have cannibalism, rub the affected area on the chick with some noxious substance such as pine tar.

At four weeks you should put roosts in the brooder set an inch or so off the floor. It could be a length of sapling or a piece of half-inch pipe. At six weeks, so long as the weather is settled, the chicks can be taken from the brooder.

There is a rule in poultry raising that says that birds of different ages should never be raised together. I have never nor will I ever consider building a separate coop for my young birds. I have always taken the young birds directly from the brooder to the larger chicken house and have never had problems. I do, however, make a small enclosure within the coop to confine the younger birds for a few days. This serves to acclimate them to their new surroundings and prevents them from wandering off or being driven away by the older birds. After a few days the young chickens can be allowed out by making a small hole or holes in their enclosure large enough to allow them passage but too small for the older birds. Now you will see what is meant by "pecking order." The older birds will chase and peck at the young ones but no damage is done. They will rarely let the younger birds eat from the feed troughs, so it is necessary to feed them in their enclosure for a few weeks until the transition is completed. It's a good idea to keep an eye on the young ones when they begin to range outside the coop so that they do not wander too far and get lost. Chickens, especially younger ones, are not noted for their sterling sense of direction once out of familiar areas. In succeeding weeks they will enlarge their world and need no watching.

Hatching Eggs

This can be the cheapest method of starting a flock if you build your own incubator (see *Appendix*) and get some fertile eggs from a neighbor. Having a broody hen do the work for you will free you from incubation or brooding problems, but you can't make a hen broody at will the way you can plug in an incubator when you want to hatch a few eggs. Details on hatching eggs can be found in the section on *Breeding* later in this chapter.

HOUSING

If you follow to a tee plans set forth in poultry books or agricultural bulletins you will end up with a lovely house that could conceivably be used for a guest house if you ever sell your birds. It will cost accordingly. Really, if you spend $50-$100 (or more!) on a coop your hens will have to lay golden eggs for your flock to be economically worthwhile. Use such plans for reference only, and with a little ingenuity and resourcefulness you'll build a good coop for a fraction of the cost.

The most important requirements for a good coop are:

Adequate floor space (at least 3 square feet per bird)

Good ventilation without drafts

Adequate lighting

Safety from predators

It is quite possible you can make use of what you have around your property. One of the beauties of chickens is that they demand very little. If you live on an old farm, chances are there is an old milk-cooling shed. In two different places we have lived we have made use of these old sheds. They convert to fine chicken coops in a few hours (at the most) and will comfortably hold up to two dozen birds. Other potential coops include old toolsheds, erstwhile outhouses, or a corner of a garage or barn. If you keep an eye out you might notice an old shed on someone else's property. Possibly the owner will part with it for free (or a modest fee), and if it's movable you're in business.

It is desirable, but not essential, to face the coop into the sun—to the south or southeast. If you are putting your chickens in a barn or other less well-lit location and are not allowing them to run free outdoors, it will be necessary to supply artificial light because chickens need 14 hours of light per day to lay optimally. Because laying hens give off a large amount of moisture it is necessary to provide good ventilation. This can simply take the form of a door that is left open (except at night) or windows that can be opened. Screened vents along the eaves of the roofline that can be closed off temporarily during very cold weather are also adequate.

It is recommended that you have one square foot of window space for every ten feet of floor space. Again, if you are making use of a preexisting structure you will have to make do with what you have. The windows in the milk shed we converted were for the most part broken, so rather than purchase new panes, I nailed chicken wire over the window and we covered it with clear plastic for protection during the winter. The chicken wire also serves to keep out the predators. Depending upon your area, predators—dogs, hawks, foxes, raccoons, and fisher cats—can lay waste to your flock unless you take measures to keep them out of the coop. Most important of all close your flock in the coop at night. Since most predators are nocturnal, we allow our chickens to range during the day, and when they go to their roosts in the evening we simply close the door.

For flooring we always use sawdust because it is cheap and easy to get. You can just as well use wood shavings, peat moss, straw, peanut hulls or ground corn cobs. Start with six to eight inches and instead of cleaning the coop out periodically let the litter build up, stirring it occasionally, taking out only the soggy spots and adding more litter every month or so. This litter, in addition to being the most sanitary and healthful, will supply nutritious feed and antibiotics for your flock because of the biological activity in the litter. Litter-raised chickens can do without animal protein or mineral supplements and such chickens are usually free from cannibalism. We usually clean out our coop once a year in the springtime. At that time we have over two feet of litter and it goes right into the garden because it is one of the best fertilizers we've ever used. After cleaning out the litter wash the coop with a disinfectant and start your litter over again.

Your coop will need roosts, located away from windows or drafts, allowing at least ten inches of space per bird.

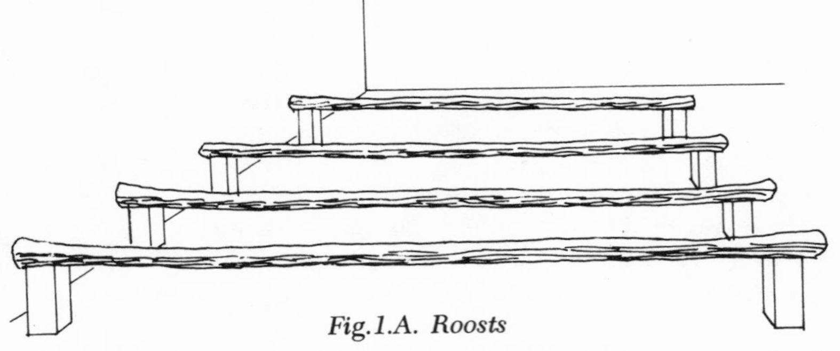

Fig.1.A. Roosts

You should set them at least two feet from the floor of the coop and space them at least a foot apart. They can be placed on the same level with each other (*Fig. 1,A*) if you have the room (this allows for better heat retention by the birds in the winter) or diagonally up the wall like treads in a stairway. (*Fig.1,B*). Two-by-two lumber is fine but straight saplings

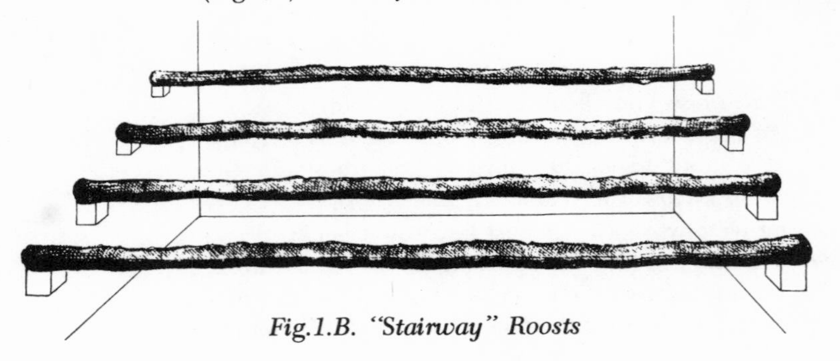

Fig.1.B. "Stairway" Roosts

about one to two inches in diameter are just as good and cost nothing.

As you will see in the section on *Feeding*, it is best to allow the birds to range as both a preventive measure against disease and for feed savings; however, if the problem of predators, lack of space, or complaining neighbors precludes ranging, it is advisable to build a run. While it's possible to keep your flock confined to the coop at all times, chickens will have more health problems and you will suffer from a mammoth feed bill. By building a run, you will allow them limited range as well as opportunities for dust baths (which control lice), not to mention fresh air and sunshine. The run should be attached to the coop so that the chickens are free to leave and enter it. The sides should be six to ten feet high to prevent your birds from "flying the coop." You can get by, however, with four-foot sides if you're housing heavy meat birds that can't get too far off the ground or if you clip a number of the primary feathers on the wings of your chickens so as to short-circuit any dreams of escape. It can be framed with two-by-fours or saplings and covered with chicken wire (so that's why it's called chicken wire!). It would be a good idea either to bury the wire in the ground or place logs around the perimeter of the run (or both) to keep predators out. As with birds that free-range, lock your birds in the coop at night.

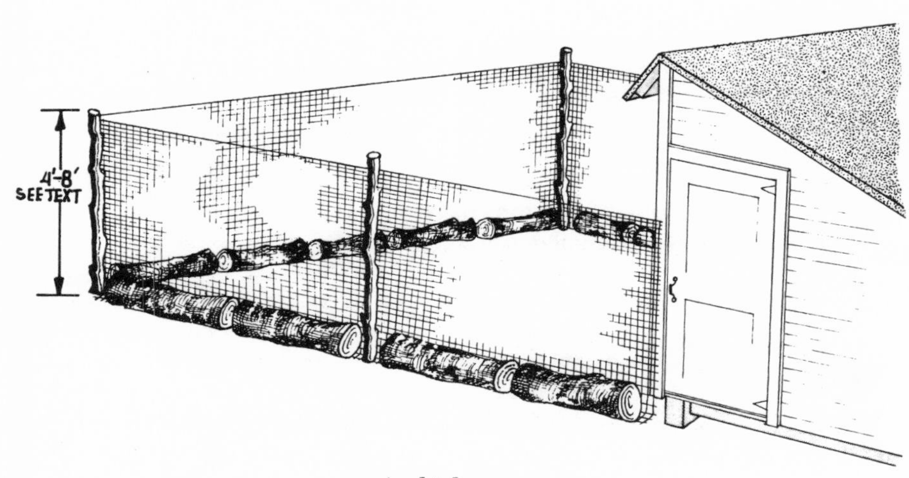

A chicken run

EQUIPMENT

You will need nest boxes at the rate of one per four layers. They should be placed in a draft-free area in the back of the coop or wherever they will be least liable to distraction from traffic in and out the door. They should be 14 inches square and a foot deep. They can be filled with wood shavings or straw and this nest material should be changed frequently to prevent soiled eggs. If space is at a premium they can be stacked as long as they are not out of reach of the hens. I have found it advisable to have a cover on the boxes since the birds seem to like it, and also to discourage roosting on the edge of the boxes with the resultant manure in the nesting material. If,

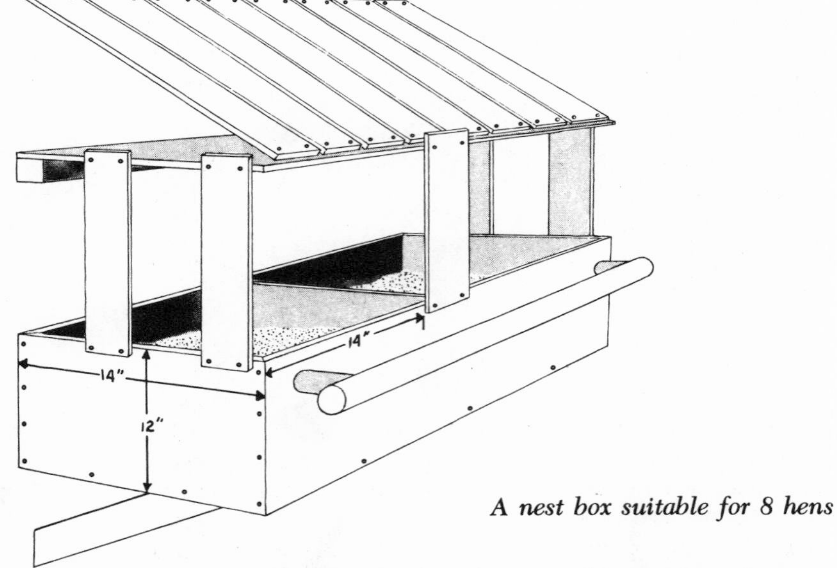

A nest box suitable for 8 hens

because of the top, the hens have trouble entering the nest, secure a small dowel in front of the box to allow the hen to fly up to it and then make her way into the nest.

You will also need feed and water dishes. They can be purchased fairly inexpensively, but so far I have resisted that urge and my homemade/makeshift feeders have done quite nicely. For feeders, the prerequisites are that the chickens be unable to scratch in them, hence wasting and contaminating feed, and that there is at least four inches of space per bird at the feeder.

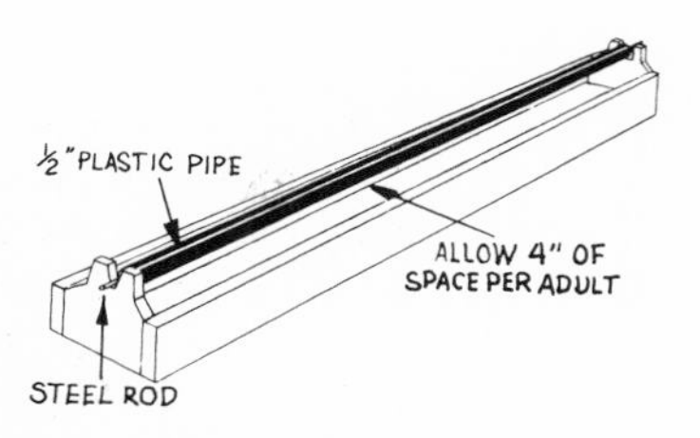

A trough-type feeder. The length of plastic pipe on the steel rod will spin if stepped on, preventing birds from roosting on it and contaminating feed.

A simple trough feeder with a reel can be built or, if scratching in the feed is a problem, covering the trough with chicken wire so only the beaks can poke through is the answer.

For a waterer I have simply used an old kitchen pan about four inches deep and about eight inches in diameter. I have made a little platform to raise it off the floor so that litter is not kicked into it. During cold weather it is essential that drinking water remain thawed or there will be a drastic reduction in egg production. I favor hanging a 60 to 75-watt light bulb over the waterer to prevent freezing. This works at the coldest Vermont temperatures, is much cheaper than an immersion water heater, and draws considerably less electricity. In a small coop this light can also be used to supply auxiliary light to maintain the 14-hour day's need for maximum production.

FEED

Healthy chickens for eggs and meat can be raised quite economically on commercial feed alone. And if you are so inclined, you can leave it at that. People wishing to push for greater self-sufficiency can make substantial savings depending upon the time they wish to spend. To avoid problems during the chicks' first six weeks of life, the most critical period, it is best to feed a commercial starter mash with a recommended 20 percent protein content. However, I have raised chicks on laying mash (16 percent protein) without any tragic results, so don't lose sleep over buying it, especially if the price difference between the two is substantial. At no other age is the furnishing of ample supplies of fresh food and water more important.

In my brooder setup, I feed and water the chicks in the end away from the light. For the first few days of their lives spread some newspaper or heavy paper over the litter and sprinkle feed on that, so the chicks learn the difference between their mash and litter. Thereafter always make clean food available free choice in chick feeders that you buy or build. (Build the same way you would for larger birds, but proportionately smaller.) I have built troughs on the side of the brooder using pieces of lath, which is even simpler and less expensive. These were two inches wide, two inches deep and as long as was needed to accommodate the chicks. The narrowness and closeness to the wall of the troughs made it difficult for the chicks to stand in them and scratch out the feed. If they do, cover the trough with chicken wire. At first allow one inch per chick of feeder space, increase to two inches at three weeks and allow three to four inches per chick at six weeks. The greatest saving you can make in feeding is never to fill their troughs more than half full with feed. If you overfill it they will waste tremendous amounts of feed (and money!).

You can buy a chick waterer, but I have always gotten by with a small bowl. I have used this method for up to 25 chicks but they do soil the water quickly, so it is essential that you check the water frequently and refill and clean the bowl if necessary. Allow ½ inch of watering space per chick.

Older Birds

You can buy a so-called commercial growing mash for chickens six weeks to laying age and a laying mash for chickens of laying age and older. Unless there is a substantial saving in growing mash (which is unlikely because of a slightly higher protein content) laying mash can be used for all chickens over six weeks of age with no ill effect. This is especially convenient if you have birds of different ages in your coop.

With older birds, as with chicks, have the mash available at all times, and don't overfill your feeders or large amounts will be lost in the litter. I have heard of buying feed in pellet form and if that is available in your locale by all means do so, as waste will be kept to an absolute minimum. If mash is spilled from a feeder, chickens will retrieve very little (if any) of it; with pellets any spilled feed will be visible and cleaned up.

To achieve optimal growth and egg production, fresh, clean water should always be available. As mentioned earlier, it is essential to keep water thawed in the winter or egg production will suffer. A bit of cider vinegar added to the water is an excellent source of minerals and is especially important for chickens that are always confined to a coop or those receiving a lot of supplemental feeding in place of commercial grain.

Ranging

I can't stress enough the importance of letting your chickens run free or "range" during the day, or at the very least, of building a run if ranging is impossible (due to predator problems or land restrictions). Commercial birds are caged (and aren't even allowed to walk, lest their leg muscles develop and get tough). They also lack many nutrients, as well as fresh air and sunshine which causes them to be the frail, disease-prone birds they are. Many poultry publications and journals spend a lot of time talking about diseases and disease losses. The fact is, I have never noticed any signs of disease in my birds, nor have I ever lost a bird to anything but a stray dog or fox. The reasons are simple: they are well-cared for, do not suffer the overcrowding and prisonlike conditions of commercial birds, and they have the fresh air, sunshine and nutritional benefits of ranging. Their legs may be a bit tougher when we eat them, but for flavor I'd take them any day over those freaks in the grocery store.

Why exactly is ranging so advantageous? First, the savings on feed is tremendous. I have read that the saving on feed averages 20 percent; but in my experience it has been much greater: 50 percent or more. I feed my flock in the morning as I free them from their night's confinement in the coop. They may make a few perfunctory pecks at the mash but are so anxious to get out the door and to the earthly delights that await them that they usually run right past me and my grain scoop. They eat grass, bugs, grit and whatever else turns a chicken on. Also, which is important for us and those of you with other stock, they clean up after the other animals. With free-ranging chickens I guarantee you will have virtually no feed waste. Our chickens clean up after our sloppy horses, grab a few mouthfuls from our obliging sow, and spend a lot of time with their favorite (and best fed) animals, the feeder pigs. We feed our feeder pigs almost totally on high-quality scraps from a local restaurant. After I empty the feed bucket into their dishes, I clean out the residue in the buckets (maybe a cup or two) with a hose and dump it on the ground. The chickens, admittedly among the dumbest barnyard creatures, have learned to come running from all corners of the farm at the sound of the hose and devour every morsel that is left. In the summer months our chickens hardly touch their mash and get along famously on animal "leftovers" and their outdoor natural diet.

Last, but far from least, ranging enables a chicken to correct its diet. A commercial mash, however complete it may look, and whatever research has gone into its development, is far from the perfect chicken food. Chickens, like most livestock, have the instinct to seek out what is lacking in the diet that man gives them. If you let your chickens range you can be confident that you are saving money, and that your flock is receiving a balanced diet. As a result they will be disease-free, healthy, wholesome and have products that reflect all these benefits both in taste and nutrition.

Scratch Feed

Scratch feed is a nonmash foodstuff that is fed in whole or cracked form by throwing it in the litter or on the ground. The formulas vary widely with some common ones listed below:

TABLE 3: Feed Formulas

Maryland		*Louisiana*		*Washington*		*Michigan*	
Corn	40%	Cracked corn	40%	Wheat	50%	Yellow corn	50%
Wheat	40%	Wheat	30%	Oats	50%	Wheat	50%
Oats	20%	Rice	30%				

You can make savings by growing your own components for one of the above mixtures or you can get by with a scratch feed of whole corn alone or corn with some sunflower seeds added. The grains should be cracked for chicks six to ten weeks old and may be fed whole thereafter. Feed lightly in the morning and more heavily at night. I recommend spreading it in their litter (remove or cover water bowl and mash feeder or they will be contaminated as the chickens scratch for feed). It will also help aerate the litter, keep it dryer and expose some of the valuable nutrients working in it.

I honestly don't use scratch feed as a regular component of my feeding program. (I get my saving by free-ranging my birds and feeding them scraps. For me, it is less time-consuming than raising my own grain for a scratch feed.) But as the table below shows it can comprise up to 50 percent of the feed program and since you can easily grow it yourself it is worthwhile considering.

TABLE 4: Suggested Feeding of Scratch Feed

Age in weeks	4	6	8	10	12	14	16
Mash (percent)	100	95	90	80	70	60	50
Scratch (percent)	0	5	10	20	30	40	50

Supplementing Commercial Feed

Three major ways of supplementing commercial feed (and saving money!): free-ranging birds, raising on built-up litter, and growing your own scratch feed, have already been considered.

You can supplement the diet of chickens confined to coops or a run with freshly cut greens (alfalfa, clover, *fresh* lawn clippings, cabbage, kale, Swiss chard and beet tops). During the winter fine legume hay, clover and alfalfa, is a good source of feed as are sprouted grains. Silages also make good supplemental feed for chickens, and their preparation and feeding is discussed in the chapter, "*Grow Your Own. . . .*"

A simple mix-it-yourself mash many of whose ingredients you can grow or procure yourself is shown in *Table 5*. A corn sheller and grain mill which can be ordered from farm catalogs (see *Appendix*) is invaluable. If you do it in large enough quantities it may be economically feasible to have a local grain mill do your grinding.

TABLE 5: A Simple Mash Mix

Ingredient	*Pounds Per*	
	100	*Ton*
Yellow corn meal	60.00	1200
Wheat middlings	15.00	300
Soybean meal (dehulled)	8.00	160
Fish meal (65% protein)	3.75	75
Mean and bone meal (47%)	1.00	20
Dried skim milk	3.00	60
Alfalfa leaf meal (20%)	2.50	50
Iodized salt	0.40	8
Ground limestone (38% calcium)	6.35	127
Totals	100.00	2000

(SOURCE: *Raising Poultry the Modern Way* by Leonard Mercia. Garden Way Publishing Co.)

If so inclined, almost anyone could raise corn, possibly wheat and soybeans and alfalfa. These constitute the bulk of the feed. Fish meal should be easy enough to make yourself (for that amount you need only be a marginal fisherman); meat and bone meal can come from the offal and bones of other livestock you butcher. If you have a dairy animal, your surplus milk can go into the feed; the ground limestone can be replaced with ground eggshells from your own chickens. Even if you have to buy some of the ingredients you can still make some savings. One note: it *is* important to grind the feed into a mash, or the chickens will tend to pick out what they like and leave the rest and not get all the benefits from the mixture.

Frankly, for the number of chickens I have, I wouldn't think of going to the trouble of mixing my own feed. But for others it may be worthwhile. I depend more on ranging, food from the litter, and having our flock clean up after our other animals. Not surprisingly this takes the least amount of my time compared to raising a number of different crops and grinding grain. I do give them any surplus milk we have that doesn't go to the pigs, and they get good-quality kitchen scraps (equal to Grade I foods from the section on *Supplementing Commercial Feed* for pigs). I don't feed them in any set proportions, but I experiment trying to find the best combination of foodstuffs I have available. With scraps watch what you feed as they are more likely to affect a day-to-day product such as eggs than a more cumulative one such as pork. In other words, if you fed your chickens a dish of garlic you would hardly want to have a conversation with one of them in a closed room, much less eat their eggs. Enough said. I have found a combination of scraps, surplus milk and ranging cuts our feed bill to practically nothing.

Whenever you are supplementing (or substituting for) commercial feed, be certain that the flock gets calcium and grit (both found in commercial feed) from other sources. The calcium (available from ground eggshells or ground oyster shells) helps form the eggshells. The grit (picked up naturally by free-ranging birds) enables the toothless chicken to grind its food. For birds not raised free range, a calcium ration and grit should be available if commercial mash is not used for feed.

MANAGEMENT

Routines

The type of "routine" a person will want to develop in managing his or her flock will take time and some mistakes will be made before it suits the particular family's needs. More than anything, experimentation is the byword.

A hen will begin laying anytime from 4½ to 6 months of age. The early walnut-sized eggs will quickly grow in size, so do not fear that you will be condemned to a year of them. As my young chickens reach maturity (also signaled by some rather discordant crowing from the young roosters), I usually butcher off all but two of my roosters, keeping a spare in case of tragedy, and keeping the *best* for future breeding, resisting the temptation of imagining how much better they would look on the dinner table.

For a hen's first year you can expect an *average* of an egg a day during her productive period (see *Molting* section). Depending upon the breed, she will lay anywhere from 200 to 240 eggs during her first year and decreasing amounts, though larger eggs, in succeeding years. You will have to decide whether to keep your hens for more than one year. After the first year, when production stops, the question is whether to keep these at a lower feed-to-egg-conversion ratio or spend the money to buy and raise new chicks to laying age. This will depend largely on the price of new chicks in your area and the amount of feed you need to purchase or whether you can supply it through your own devices.

If you wish to keep your hens another year you will have to accept the molt with understanding . . . and no eggs. I don't enjoy buying eggs, so I have worked out a routine that enables me to "get around" molting economically. We started our flock with pullets and after a summer of production, a molt and a practically eggless winter we hatched out 25 chicks the following spring. By the time the original birds went into molt again the next fall, the chicks had begun laying. We butchered the older birds and took the new layers through the winter (allowing 14 hours of light and plenty of thawed water so as to not upset production) without a molt and hatched another batch of chicks that spring. Again, we butchered the older hens that fall as the new chicks began laying. In this way we never had our hens go through molt, had plenty of meat, and never went through gaps in egg production.

Molting

At the end of a laying year (or at times due to disease or mismanagement), a hen molts, or drops its feathers, in order to renew them. They look like hell, *don't lay* (some do but these are rare), and to top it all off, tend to eat more. Depending on the bird, the molt cycle (dropping and growing feathers) can take from 10 to 24 weeks, with little or no egg production during that time. Naturally, those taking longer to molt should be culled from the flock in favor of fast molters. Fast molters often lay for a few weeks after dropping feathers, so egg loss is at a minimum. They tend to drop feathers in groups rather than singly and grow them back in groups, hence returning to production more quickly. Slow molters drop one feather at a time and renew them one at a time. They often begin molting earlier and continue for months. Eat these birds.

When molting begins feathers are shed in the following order: head, neck, breast, body, wings and tail. Some birds will molt after only eight or nine months of production (early molters—cull them), while others will lay for 12 months before the onset of molt. The primary feathers of the wing (see *Fig.* 5) are used to chart the molt because they are dropped and renewed in a set order. By examing the primary feathers one

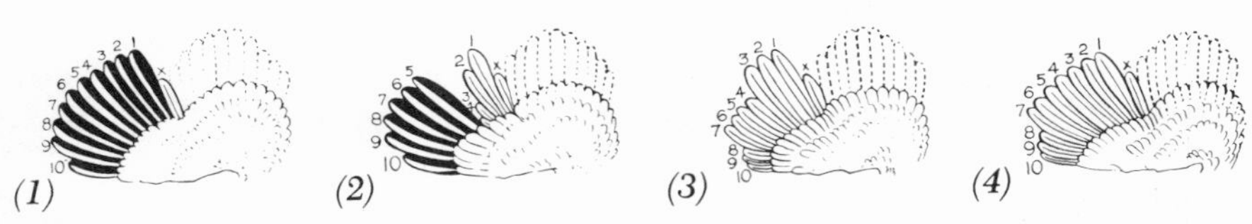

(Fig. 5) Wings during different stages of molt. (1) shows the 10 old primary feathers (black), and the secondary feathers (broken outline), separated by the axial feather (x). (2) shows a slow molter at six weeks of molt, with one fully grown primary and feathers 2,3, and 4 developing at two-week intervals. In contrast, (3), a fast molter, has all new feathers. Feathers 1 to 3 were dropped first (now fully developed); feathers 4 to 7 were dropped next (now four weeks old); and feathers 8 to 10 were dropped last (now two weeks old). Two weeks later (4), feathers 1 to 7 fully grown. Fast molt took 10 weeks, compared to 24 weeks for slow molt. See text for futher explanation. (South Dakota Extension drawing.)

can estimate how long a bird has been in molt and how soon before it will return to laying. If the feathers are dropped singly or in groups, one can decide if the bird is a slow or fast molter. The first primary feather to be dropped is in the middle of the wing (Number 1 in the upper left corner of *Fig.* 5) and it takes six weeks for a new primary to be fully regrown. Normally, the next primary drops in two weeks; the next, two weeks after that and so on. When a fast molter drops two or more primaries at a time they will also grow out together. By dropping primaries in groups of two or more, these fast molters (and high producers) complete a molt more rapidly.

Force Molting

For birds that do not normally go into molt after 12 months, or for those birds whose production lags prior to molting, it may be desirable to use what is known as force molting. Through this procedure the birds will drop their feathers and will return into production after new feathers are grown, which may be as early as eight weeks. Egg production, as it normally would be in the second year, is lower after force molting. In addition, the eggs are larger, feed consumption is higher and the health of the flock will be poorer. I have never used it because I, unlike a commercial poultryman, do not live or die with relatively minute changes in production, and I also consider it an unnatural strain on a bird. It is, however, an alternative to buying a new set of birds after the first year of production.

TABLE 6: **Force Molting Procedure**

(A) *For normal weather conditions:*	
Day 1	Decrease light to 8 hours per day in light-tight houses or no-natural-light in window houses. Remove all feed and water.
Day 3	Provide water.
Day 8	Provide growing mash—40% normal consumption level.
Day 22	Restore lights to premolt level and full-feed laying ration.
(B) *For use during hot weather:*	
Day 1	Decrease light as above. Remove feed but allow water free choice.
Day 8	Provide growing mash—40% normal consumption level.
Day 22	Restore lights to premolt level and full-feed laying ration.

(SOURCE: *Raising Poultry the Modern Way* by Leonard Mercia. Garden Way Publishing Co.)

Culling

Culling is the removal of unhealthy or poor producers from a flock. Unless diseased, a cull bird offers the compensation that it is at least as good for the table (maybe meatier due to lack of strain because of egg producing) as a layer. Eat well.

Egg Production

As mentioned earlier, maximum egg production cannot be maintained without thawed water in the winter or 14 hours of light per day. As the days grow shorter artificial light must be supplied so that the 14-hour requirement is met. Light can be turned on in the morning before dawn or the evening after dusk or in combination, as long as 14 hours are provided. Since I will not buy an electric timer to control my lights (worth about 30 dozen eggs), I have found putting the light on at dusk and turning it off before bedtime most convenient for me. Suit yourself.

Usually you can get by collecting eggs once a day, but especially during cold or warm weather more frequent collection will be necessary to insure the high quality of your eggs. Free-ranging birds can present a problem by laying some of their eggs out of the nest. If production drops suddenly when the hen isn't in molt, this may be the cause. Often it is due to a filthy nest box, but more often it is the capriciousness of the birds. If I can find out where they're laying (try following them for a while; they love tall haystacks and similar out-of-the-way places), and can collect them easily, I usually leave them be. If not, try confining them to the coop for a few days so they have no choice.

If a hen becomes broody (i.e., wants to hatch out eggs) when you don't want her to, a broody coop suspended from the ceiling of the coop will cure her. A broody coop should be small with a slatted bottom and plenty of food and water should be provided. In three or four days she'll get the message and she can be freed.

Caponizing

Caponizing, or the rendering of the male chicken a castrato (as far as my research tells me they are not yet organized into choir groups), either surgically or chemically, is beyond the scope of this book. Leonard Mercia's book, *Raising Poultry The Modern Way* (see *Appendix*), goes into considerable detail about this subject. If you want to raise capons, and they make darned good eating, they can be bought from most local poultry farms. The cost is about four to five times that of day-old chicks, and they are sold at about four to six weeks.

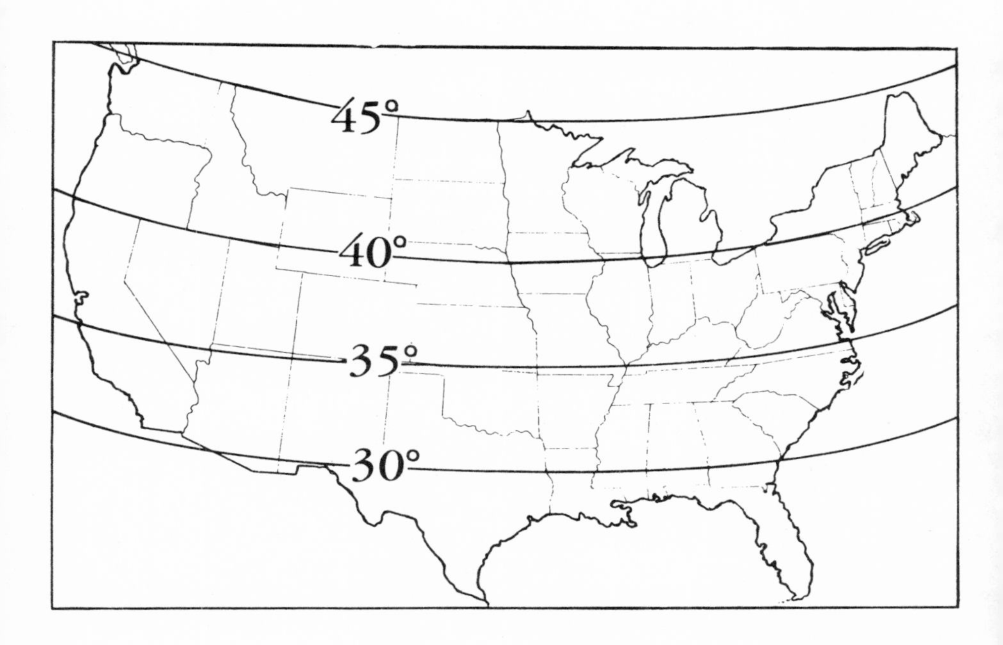

Hours of Daylight By Month

Month	45° N. Lat.	40° N. Lat.	35° N. Lat.	30° N. Lat.
January	*9:09*	*9:39*	*10:04*	*10:25*
February	*10:26*	*10:41*	*10:56*	*11:09*
March	*11:53*	*11:54*	*11:57*	*11:58*
April	*13:29*	*13:15*	*13:04*	*12:53*
May	*14:51*	*14:23*	*13:59*	*13:39*
June	*15:35*	*15:00*	*14:30*	*14:04*
July	*15:07*	*14:34*	*14:07*	*13:44*
August	*14:06*	*13:46*	*13:29*	*13:14*
September	*12:33*	*12:30*	*12:26*	*12:22*
October	*11:01*	*11:12*	*11:20*	*11:28*
November	*9:34*	*9:59*	*10:21*	*10:41*
December	*8:48*	*9:22*	*9:50*	*10:13*

You must supplement normal daylight to maintain 14 hours of light per day to obtain maximum egg production. Refer to the map and the chart to determine how many hours of supplementary light (if any) you must supply in your state. More accurate readings may be gained through interpolation.

Handling

The best way to carry a chicken is to hold it upside down by the feet. The blood rushing to its head will subdue it after a few moments of indignation. Unless you lock up your chickens at night and can corner them in the coop, catching a chicken is another story. I usually wait for them to put themselves to bed at night or grab one before letting them out in the morning. I have, inadvertently, caught one in a Havahart trap when I was trying to catch rats. Poultry hooks (that snag a leg), available from most farm catalogs (see *Appendix*), make what is potentially a day-long task a snap.

Predators

Most predators attack during the night, so that is when it's important to lock up your chickens in a well-protected coop. Hawks and dogs are the most frequent daylight predators. However, we have never lost a chicken to anything except *our dog.* Hawks will swoop out of the sky and make off with peeping chicks or young birds or pick apart on the ground those too heavy to carry off. If you are bothered by them, keep your young ones in until they are larger or set a large animal trap baited with a dead chicken *out of the reach of children and pets.*

Dogs are another matter; chicken chasing is a top sport. More often than not most of your losses will come from birds that are scared away and lost (and subsequently devoured by other predators), rather than those which are actually killed by the dog. If you can't catch the culprit, get rid of it, or persuade a neighbor to tie it, you will have, regrettably, to build a run. If you have only one persistent offender, try bringing him to the coop and beating him severely with a dead chicken. Some people say once a chicken killer, always one, but I don't buy it. Our dog once went on a spree and killed or scared away 17 two-month-pullets, but punishing him and keeping him away from the coop has cured him of looking even cross-eyed at a chicken ever since. Another method that we used was to tie the dead bird around the dog's neck but close enough to his body to prevent him from tearing it free. Tie him away from the house for however many days it takes for the chicken to smell to high heaven and fall off. As a last resort—I've never tried it—you can enclose the dog and dead chicken in a barrel and give the barrel to a goat for a few days.

BREEDING

While you can get eggs with a flock of hens alone, to get fertile eggs you need a rooster. Hatching your own eggs enables you to self-generate replacements and cut your ties to hatcheries or poultrymen. One rooster can handle up to 15 hens and have the time of his life. (Note: 14 hours of light a day is also essential for maximum fertility of your rooster.) It may be deduced that human males learned male chauvinism from the rooster. As a rooster is routinely pecking about the yard he will, as if a button on a distant control board was pushed, suddenly look up, begin chasing a hen, collar her by the neck, drive her face into the ground, and mount her. It is hardly the tenderest of nature's mating rituals.

Artificial Incubation

Incubators can be purchased from feed stores, Sears Roebuck, or ordered from farm catalogs; but unless you plan to hatch eggs on a large scale, I doubt the economic value of such an operation (and economics, presumably is why you'd choose to hatch eggs in the first place). It is possible for anyone to construct an incubator with a minimum of time and money (see *Appendix*).

If you are collecting your fertile eggs day by day, you must keep them under carefully controlled conditions until you incubate them. Storage of up to a week has a minimal effect on an egg's hatchability but after 12 days the viability declines quickly. Eggs must be kept at 55°-65° at a humidity of 70 to 80 percent. Under these conditions the egg cell remains dormant. At 80°-90° the cell germinates but soon dies; at temperature above 110° the cell will also die. Eggs should be stored, small end down, at an angle of 35° and rotated twice each day. Select only well-formed eggs and discard any that are misshapen, cracked, or unusually large or small.

An incubator must be set at a temperature within the very strict limits of 99°-105°. Start the incubator the day before to check its operation and to allow it to heat up before inserting the eggs. Warm the eggs up to room temperature before placing them in the incubator so the temperature will not fall sharply when they are put in. It is necessary to have a small pan of water on the floor of the incubator to prevent the air from becoming too dry. Finally, it is essential to turn the eggs three to four times a day. Make a small mark on one side of the egg to avoid confusion when turning.

If you have done everything correctly, in 21 days (give or take a day or two either way) you should notice cracks appearing on the eggs and even hear little peeps coming from them. The breaking-out process is a long and arduous one for the chick and can take up to a day. If after a long time a chick still hasn't emerged from the egg, it is generally recommended that you leave it and let nature take its course; however, I can never bear to see a beak poking out and leave the chick to die, so I usually help the little fellow out. Thus far I have yet to have any such chicks die and my henhouse has not been struck by lightning from above.

In planning the ultimate size of your flock bank on an average hatch of 60 percent. I usually raise my chickens for both eggs and meat so I am not unduly concerned with the sex of the chicks. I just butcher off most of the cockerels as they come of age. If you are interested in raising only laying birds and wish to do away with cockerels rather than pour feed into them until they start crowing, it is possible to sex day-old chicks. It is a tricky process and it is too involved to go into here, but there is a good article devoted to it in the *Mother Earth News* (Issue No. 27, May 1974, p. 60). A method of determining the sex of the chick prior to hatching by the shape of the egg is disputed by some, but I feel there is some merit to it. (Anyway, there's no harm in trying.) Simply: those eggs that are most pointed are cockerels; those with most oval are pullets. Take only the most pronounced in each class and discard those falling in between. There are some breeds, so-called sex-links, whereby the sex of the chick is distinguishable by color. Because of the multitude of breeds it is best to ask a poultryman or extension agent for your particular breed.

Once a chick hatches keep it in the incubator until it is dry and fluffy. From then on treat it as you would a day-old chick that you've purchased.

Natural Incubation

I find this method much easier because the mother hen will do the brooding for you and you don't have to bother building an incubator. On the other hand, a hen will not just put herself to hatching eggs whenever you want her to. On the whole it is less efficient, and it puts a bird out of production for the hatching and brooding period.

All breeds are not equally suited for mothering. The more nervous, finely bred birds (usually egg birds) have had it pretty effectively bred out of them. Bantams, Wyandottes, Plymouth Rocks and Rhode Island Reds, however, make good mothers, and will generally set, especially as they get older. If you want chicks, keep an eye out for a broody hen and place some eggs under her if she doesn't have enough of her own. A large hen can cover up to 15 eggs and a medium one around 9 or 10. Make sure she isn't disturbed too much and have water and feed nearby for her infrequent feedings. I will leave my setting hens in the coop, rather than moving them, as unfamiliar surroundings usually break them of their broodiness. If you notice other hens crowding her out of the box and laying too many eggs with her, you probably don't have enough nest boxes or the others are soiled. Once she sets she will take over from there. She naturally turns the eggs every day and when they hatch she will brood them until they are old enough to go out on their own. Often there is a problem of acceptance of the mother and her clutch by the other birds, but normally after a few weeks they will become accepted. At times we've had the problem of young chickens becoming wild, roosting in trees and returning to the coop only for food. In this case, simply lock them in the coop for a week and they should be broken of this.

HEALTH

In combating diseases, again your most effective tack is prevention. Housing your birds on built-up litter will cut down on disease because of the production of antibiotics in the litter. Letting your chickens range free to correct their diet and reap the benefits of fresh air and sunshine is the best way to promote good health. Disease with my flock, and most small flocks, should seldom be a problem. Disease is more likely to occur in crowded, confined commercial flocks which rarely see the light of day. A general sign of ailing birds will be a drop in feed and water consumption, lower egg production, listlessness and death. Unless you are going into large-scale production, the purchase of a veterinary book for chickens only is unnecessary. The disease table listed in the *Appendix* should tell you what you need to know.

BUTCHERING

Chickens are good to start your livestock farming on, and they are relatively easy to butcher; quick, not messy and since it is very hard to form a strong bond of friendship with a chicken, relatively easy to dispatch.

Whenever you choose to butcher your bird you should deprive it of food (but not water) for the preceding day.

TABLE 7: Butchering Terminology for Chickens

Term	*Sex*	*Weight or Age*	
Broiler	Either	2½ lbs	Less than 8 months old
Fryer	Either	2½-3½ lbs.	
Roaster	Either	3½-5 lbs.	
Capon	Castrated male	6-8 lbs.	
Fowl	Female	Over 8 months	
Cock or stag	Male	Over 8 months	

When capturing a hen for eventual butchering, be watchful of any roosters you might have, as they are never pleased to lose a member of their harem and can get quite nasty. Our present rooster clucks indignantly as I stalk one of his lovelies and then attacks me after the catch, but only after I turn my back. I can usually outrun him.

Whatever your choice of butchering size and age (and whatever the personality of your rooster), you will have two methods of killing (a third, breaking the neck, is rarely used and is excluded): the most well known (1) beheading, and (2) sticking and cutting the throat. Both are equally effective and perhaps beheading is best for the beginner since it is the quickest and surest. Holding the bird by the legs, I tie some bailing twine around its feet so that it can hang for bleeding after I've cut its head off. (By hanging it this way you will also be spared the anguish of your first chicken flapping madly about the yard without a head.) Have a sharp ax and a chopping block (a stump will do) with two nails driven in (but sticking up two inches) parallel to each other and about one inch apart. Insert the chicken's head between the two nails and pull gently. The beak will get caught between the nails and the neck will stretch giving you a surer shot. After chopping the head off hang it up immediately and let it bleed until dry. I usually let it hang ten minutes or so; much longer, especially in colder weather, will make the bird harder to pluck.

When sticking, also tie the bird's feet and then hang it upside down. Take a sharp, short-bladed knife and cut the throat at the base of the neck, being sure you sever the main

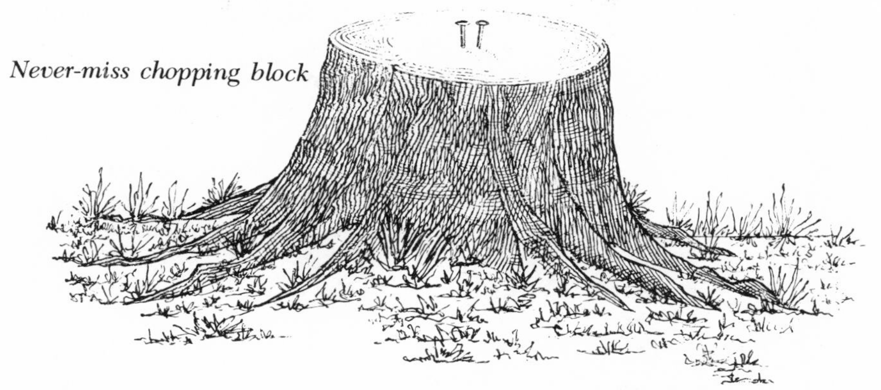
Never-miss chopping block

arteries. Immediately insert the point of the knife into the bird's mouth and force it through the roof toward the back of the head into the brain cavity and give the knife a quarter turn. The piercing of the brain causes a squawk and a convulsive wing flapping that loosens the feathers and makes the plucking easier.

Plucking

Plucking may be done by the wet or dry method. Dry plucking is begun immediately after the bird is bled and before it becomes cool, since this tends to set the feathers. This is best done after sticking (rather than beheading). Wet plucking is accomplished after dipping the bird in hot (150°-190°F.) water for a half minute to a minute. Too long and the skin will cook. If you wish to freeze your bird for any length of time, it is best to dry pluck since the scalding tends to cut down on freezer life.

Plucking is hardly fun, and after doing a few birds you may be ready to switch to rabbits for your source of this type of meat and just raise chickens for eggs. Anyway, your skill and speed at plucking will improve with time. It is best to take your time at first, lest you tear the skin and damage the bird. The best order of plucking is as follows: wings, breast, body, back, legs and neck. If any small pinfeathers are left, they may be removed by using a dull knife or by brushing the bird with a mixture of six ounces of melted paraffin and seven quarts boiling water. After it has hardened pull it and the feathers off. The mixture can be reused by straining it through the cheesecloth and cooling. Small feathers and hair may be singed off by using a gas flame or a candle.

Dressing It is best to work with a chilled carcass. A sharp knife is essential and poultry shears are helpful but not necessary. First cut off the head and, peeling back the skin, sever the neck and save it for the stockpot. Then remove the feet at the hock joints (after washing they make an excellent addition to the stockpot). Make a cut in the vent large enough to accommodate the hand. If the bird is of laying age you may find a fully formed egg and see a string of yolks in varying stages of development. Pull out the innards, saving the heart, liver and gizzard. The rest of the entrails may be fed to cats (never feed to dogs or they may get the "taste" and set out to catch their own), or the pig, or chopped and fed as an addition/substitute to chicken feed. Carefully remove the gallbladder from the liver and remove the inner sac from the gizzard. Push your fingers in the front and pull out any stray parts. Flush with water and make sure all pieces of blood clots are removed. After washing the bird thoroughly, it may be roasted, or cut up for frying or broiling or frozen as is.

II. DUCKS

Ducks are hardier than chickens. They require less attention and they grow more quickly. While they do not forage for food as much as geese, they will still augment their diet by ranging, and they will produce tasty meat very economically in only eight to ten weeks.

They need very little space and almost no housing. Your ducks will not require a pond or a brook, but if you have one they will be eternally grateful. The meat is delicious, not at all gamy like wild duck and with less fat. You also have the opportunity to gather duck eggs and even some duck down for pillows or whatever else you may choose to do with it.

Ducks have something else over chickens. They have personalities! Few people get pleasure from watching chickens, as they have the maddening tendency to behave the same way all the time. Ducks frolicking on a pond or waddling mechanically in a line to some distant objective are a sight to behold. Be careful: you can make ducks into pets (try that with chickens), and if their ultimate destination is your dinner table this can put a crimp in your plans.

BREEDS

There are some breeds of ducks that are well suited to meat production and others whose forte is egg production. Unfortunately, unlike chickens, there is no such thing as a dual-purpose breed of duck. You can eat good egg layers, but the carcass will be small and of poorer quality, since most of their energy goes into the production of eggs, not meat. The so-called meat birds will lay poorly, however, and will supply you with a trickle of eggs along with meat. Duck eggs, while larger than chicken eggs, are good for baking, but many people find them unsuitable for "straight" eating. They are said to be higher in protein and lower in cholesterol than chicken eggs. If unfamiliar with their taste, find a neighbor or a local hatchery that has some surplus eggs and try them before you decide to live on them.

Meat Breeds

White Pekin: This is the Long Island duck you've heard so much about and is the only breed raised commercially in the United States. They are large birds with white plumage, orange-yellow bills, reddish-yellow legs and feet, and yellow skin. Adult drakes reach a weight of about nine pounds while adult ducks (the females) reach about eight pounds. They produce excellent meat and reach market weight of seven pounds in from eight to ten weeks. They are also reasonably good egg producers, a duck laying an average of 160 eggs per year. They are high-strung birds and as a result poor setters. If you wish to hatch their eggs, it must be done by artificial incubation or by enlisting the services of a broody hen.

White Pekin

Muscovy: While many varieties of this native of South America exist, the white ones (because they dress out more attractively) are most desirable for meat production. They have white skin and the darkening around the eyes resembles a mask. Adult drakes weigh ten pounds and ducks weigh about seven. They reach market weight in about 10 to 17 weeks. They are poor egg producers (average 40 to 45 eggs per year) but they are the best setters of all meat producers.

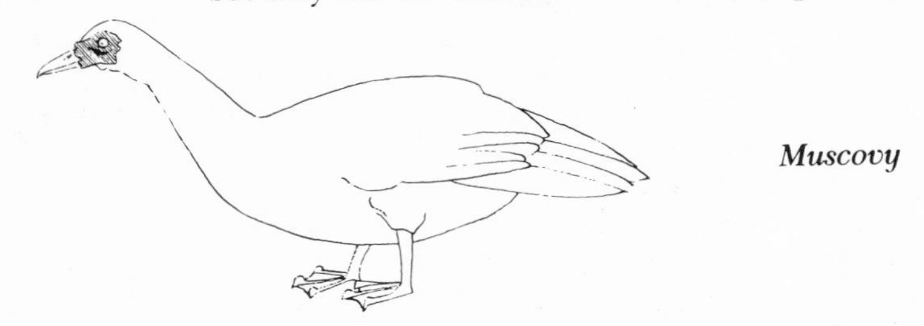

Muscovy

Aylesbury: This is the English counterpart of our White Pekin and its quack has a distinctly British air. Like the Pekin, they reach meat weight (seven pounds) in eight to ten weeks. They have white feathers, white skin, flesh-colored bills and light orange legs and feet. Adult drakes weigh nine pounds, ducks eight. While they are not as nervous as Pekins, they also do not have much interest in setting. They are not quite so prolific in egg production as Pekins.

Other meat breeds include Rouen, Cayuga, Call and Swedish.

Egg Breeds

Khaki Campbell: Khakis are able to surpass even the highest producing chickens in egg production. They are, however, not valued for their meat since drakes and ducks reach a mature weight of only 4½ pounds.

The Khaki drakes are bronze on their lower back, tail coverts, heads and necks, the rest being khaki-colored. Their bills are green and they have orange legs and feet. The females have seal-brown heads and necks; the rest is khaki-colored. Their bills are greenish-black and their legs and feet are brown.

Indian Runner: Indian Runners fall short of Khakis in egg production, but they are still good producers in their own right. There are three types: White, Penciled, and Fawn and White. All three have orange to reddish-orange feet and legs. They stand quite erect, their carriage being almost perpendicular to the ground. Their adult weight is similar to that of Khakis and they are likewise not valued for their meat.

Indian Runner

PURCHASE

As with chickens, you have three choices: (1) hatching eggs, (2) buying day-old ducklings, or (3) buying older stock for breeding. Your choice will obviously depend on how you will raise your birds. If you plan to raise only a few ducks for meat, buy a few day-old ducklings from a local hatchery or friend and raise them. This frees you from having to have incubating equipment and fussing with breeding and hatching. If a larger flock is desired, you can buy fertile duck eggs or buy a lot of ducklings and go from there, selecting the best for future breeding stock.

Hatching

In incubating domestic duck eggs the hatching time is 28 days for all breeds except Muscovy which is 35. You can use the same incubating and brooding methods as with hatching chickens (see chapter on *Chickens* and *Appendix* on incubators). The eggs should be placed in the incubator small end down, or if that is impossible rest them on their sides.

For natural hatching, Muscovys are the best breed to use and frequently the only ducks that will hatch eggs. If you don't have a setting duck, you can make use of the bane of the barnyard, the broody hen to do your hatching. A broody hen will handle up to ten duck eggs (don't let her take more than she can handle since this will affect the hatchability of all of them), a duck slighty more. Sprinkle warm water on the eggs each day if you are using a hen or a duck that has no access to water, since duck eggs need more moisture than chicken eggs. When they hatch, in the case of a duck mother, let nature take its course; in the case of a hen foster mother, confine her and her brood of ugly chicklings so that she cannot wander off, as they will not be able to keep up with her the way chicks do, and they may get lost. After four weeks they should be able to keep up with her and she can be set free.

Day-Old Ducklings

In purchasing day-old ducklings locally, you have the advantage of being able to buy fewer (mail-order minimums are often 25), but you may not be able to have your choice of breed. In any event, you will need a brooder. Handle brooding as with chicks but allow one square foot of space per bird. Ducklings normally need brooding for only four weeks and may even get by with two or three weeks. In warmer summer weather a light bulb in their shelter may suffice. Do not allow a duckling to get soaked by rain before four weeks of age or until it is well feathered. It should also not be allowed to swim before six weeks because the oil is not fully distributed on its feathers, and it will be more susceptible to chills and sickness. Shade must also be available to young birds.

Older Stock

This applies to choice of mature stock which are bought for breeding or to picking potential breeders from your own flock. At six to seven weeks drakes and ducks can be distinguished not only by the difference in coloration (males have brighter plumage), but also by their sounds. Females "honk" and drakes "belch." You will require one drake for every six females for breeding purposes. Look for ducks with evenly colored feet, legs and bills. Choose birds that are vigorous and alert and are heavy and solid, with broad breasts and necks that are not too long. Plumage should be even and glossy. Find out, if you can, the breeding records of the parents: choose ducks from parents that have exhibited high fertility, hatchability and egg production.

Predators

Predators are generally the same as for chickens. Your dog, while perhaps accustomed to chickens, may see the newcomers as fair game. Watch him.

HOUSING

If you plan to raise ducks for meat in the summer months, housing is quite simple. If you don't have a handy pond or stream, you could just as easily house them with the chickens—if they will stand for it. If the chickens peck and chase them, as they are wont to do, make a little pen within the coop that only the ducklings can scoot under. Within a few weeks they will probably be accepted, but if not, they will be close to butchering age anyway and out of your and your chickens' hair (or feathers).

Allow ducks to range free all day and simply lure them with grain into a small enclosure and lock them in at night to safeguard them from predators and hungry neighbors. A simple, movable hutch should have a dry floor and furnish protection from severe weather and predators (see *Fig. 8*). They don't need roosts, but allow five to six square feet of floor space per bird.

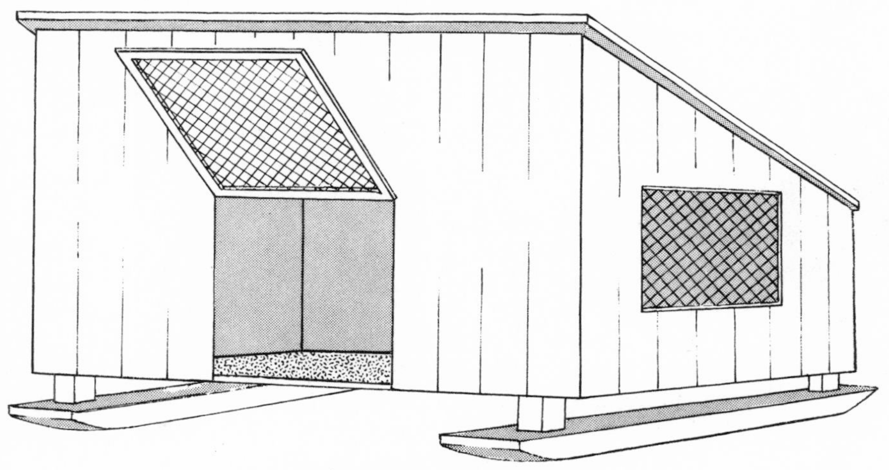

A movable house for ducks or geese

If you wish to go into breeding, you will need a more permanent house. It can be much the same design as a chicken coop, but no roosts are needed. Ducks are more resistant to cold than chickens, so you don't need to be so careful about plugging up drafts in the building. However, you shouldn't allow any snow or rain to seep in, and it should, of course, be predator-proof. Space requirements are the same as for the temporary shelter described above. If feather pulling among the ducks is a problem, they may be overcrowded. Give them more space. If this doesn't work, debill them (clip off the front part of the upper bill). Keep the house scrupulously clean. A built-up litter system as with chickens works well, and be sure to shovel out any wet spots.

If you allow your birds to range, as I would strongly suggest, they will be healthy and disease-resistant. Keep them away from pools of stagnant water, since these are havens for disease. If for any number of reasons (limited space, predators, large flocks, etc.) you must fence your birds in order to let them run, allow at least 40 square feet per bird. Keep the run very clean. Because ducks can be quite messy, sand is the best ground cover and it helps if you situate the run so that it slants downhill away from the house. For run-confined birds, you must throw in fresh greens daily to realize any feed economy.

EQUIPMENT

Ducks do not need swimming water to live, nor is it necessary for fertilized eggs, but if you have a pond your ducks will love you for it. Don't allow the population to get too high or your pond will be a mess and will smell awful. For ducks confined to a run or on a pondless farm, you can supply them with a pool they can splash in and clean themselves. This is especially good if you have ducks hatching eggs, since it will enable the duck to supply more moisture to them. The pool should measure at least three feet by one foot and be deep enough so the ducks can totally immerse themselves. It can be constructed out of concrete or wood, or you can use an old feed trough or discarded bathtub. Be sure to change the water frequently to prevent stagnation.

Feed Dishes

Your ducklings can be fed from chick feeders or shallow troughs or pans. Since they do not scratch their food as chickens do, waste is less of a problem. As they get older, you can feed them from larger troughs or from a self-feeding hopper (see *Fig.* 9). This hopper is very useful if your birds are

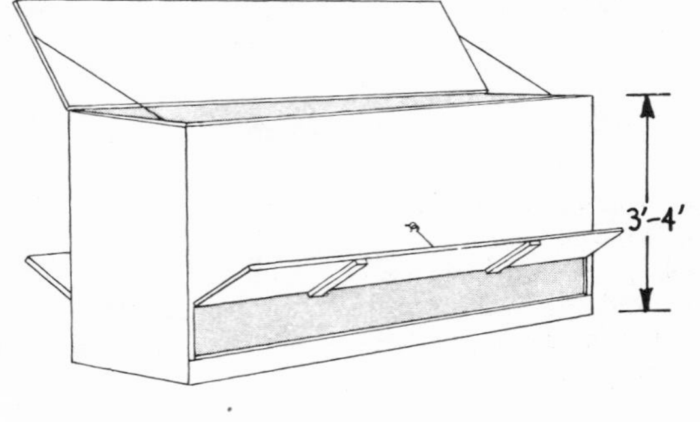

A self-feeding range feeder for ducks or geese. Lower panels may be closed at night to prevent piracy by other poultry or rodents.

free-ranging, since it obviates your having to make a larger house to hold the feeding equipment too. If you have free-ranging chickens, food piracy may be a problem. We have handled this by separating them from each others' ranging area.

Waterers

The duck's waterer must be constructed so that the whole bill (but not the whole body) can be immersed and the nostrils can be cleaned. Ducks are very messy drinkers so if they are watered inside their house, place the waterer on wire off the ground and have a lot of absorbent litter underneath. Better yet, don't give them their water while they're cooped up. They can adjust to being without it while closed in at night and their egg production and growth will not be affected. If you do keep the waterer outside, the food must be kept out of the coop too, since a duck eating dry food without access to water may strangle.

Nest Boxes

If you plan to keep your ducks for egg production or breeding, nest boxes must be provided. They can be constructed like chickens' nest boxes, but should not have tops. They should be a bit larger, at least 15 inches square and 18 inches deep, and they should be set on the ground. Nest material can be the same as for chickens.

Stampede Lights

When startled at night, ducks have a tendency to run in circles, possibly causing injury or death. A low-wattage night light will keep this from happening.

FEED

Ducks benefit (or maybe it should be said that *you* benefit) from an excellent feed conversion ratio (3:1), and this should enable you to raise meat and eggs economically. While ducks are not so good at foraging for food as geese, they will still augment their diet if they are allowed free range, or supplied with greens if penned.

The following table outlines the rations that should be fed to ducks:

TABLE 8: Feeding Rations for Ducks

Starter for the first two weeks (20%-22% protein)
Grower for meat birds, 2 weeks to market (15%-18% protein)
Breeder developer for breeding and egg flock, six to seven weeks of age until one month before production
Breeder from one month prior to production

Pelleted feed, if available, is preferred to mashes, as there is less waste and the ducks convert it to meat at a better rate. Most feed stores should carry duck feed or feed that is suitable for them. Chicken feed can be substituted for *growing* flocks, but *medications present in chick starter can kill young ducklings.* They will have to be fed either an unmedicated chick starter or feed formulated for ducklings. The starter feed can be in the form of pellets (no larger than 1/8-inch) or mash that is slightly wet. Allow the feed to be consumed free choice and, of course, supply fresh water.

For meat ducks, allow free choice all the grower feed they will eat. Again, mash can be utilized but a 3/16-inch pellet will be utilized with greater efficiency and less waste. For the ducks you plan to use for breeding you should feed the breeder developer ration. To prevent them from becoming too fat, feed this on a limited basis—1/2 pound per duck per day, split between morning and evening feedings.

At one month prior to egg production (explained in section on *Breeding*), switch the ducks from the breeder developer ration to a breeder ration and feed them free choice. If feed does not contain a calcium supplement (which is unlikely if it's a commercial feed), ground eggshells, chicken or duck, should be added.

Supplementing Commercial Feed

While their feed conversion ability is good, and their relatively small level of feed consumption makes supplemental feed less important than for pigs, you can still make some savings. It is impractical to mix your own feed, but you can augment their diet by allowing them to free-range. They will obtain a great deal of their feed requirements naturally and balance their diet at the same time. Ducks that are penned should be supplied with *fresh* greens at least once daily.

BREEDING

Ducks that have hatched from April through July will reach maturity at seven months. Those hatched between September and January will mature in five or six months because of the increasing length of the days in their growing period. It is best to pick your future breeding stock when they are six to seven weeks of age and begin feeding them as indicated in the *Feed* section. (The sexing of ducks is discussed in the *Purchase* section and under *Breeding* in Part III, *Geese,* that follows.)

You will need one drake for every six ducks, but I suggest at least one extra in the event that one of the drakes meets an untimely end.

When your birds are near maturity, and three weeks before you wish them to begin full egg production, increase the light in their coop so that they receive 14 hours of light a day. (This is discussed at some length in the chapter on chickens.) Give the drakes an extra one or two weeks' head start so you'll be sure they're ready for the ducks. Bringing females into full production (i.e., supplying them with 14 hours of light) is not recommended before seven months of age because they will tend to produce smaller eggs with a lower hatchability.

Once the duck begins laying, it will reach its production peak within five to six weeks. If you are interested in hatching eggs, the fertility is highest when egg production is highest, and the greatest hatchability will occur after the first few settings (probably because the birds will have improved mating proficiency). The details of handling fertile eggs, incubation and brooding are given in earlier sections.

Ducks lay most of their eggs before 7:00 A.M., so keeping them locked up until after they are through laying will not markedly affect their feed consumption (if you keep the feed outside), or their ability to range. The number of eggs they lay will, of course, depend upon the breed. High production will be maintained for five or six months and then will taper off gradually, allowing the ducks a rest period.

HEALTH

At the risk of being redundant, prevention is the byword. Clean pens and runs frequently and avoid crowding and you shouldn't even need to consult the disease chart in the *Appendix*. Keep your young ducks from getting chilled and keep all ducks away from stagnant water.

BUTCHERING

Ducks reach market weight at seven to ten weeks of age (except Muscovy: 10 to 17 weeks). Butcher those with firm, plump carcasses without many pinfeathers. Raising them beyond the recommended ages and weights will perhaps produce larger ducks, but not so economically.

As with other stock, keep ducks from food for 12 hours before slaughter but allow access to water. Methods of killing are the same as for chickens. You can dry-pluck or scald; wax on the pinfeathers may be quite helpful. Butcher as with chickens.

III. GEESE

Geese are the hardiest of all poultry; they are the most disease-resistant and the cheapest to raise. Because they are such good foragers, and can be set out to forage when they are young, they can be raised with an absolute minimum of feed. They also make good "watchdogs," a plus in crime-ridden suburbs. Walk a goose every night, and see if you get mugged. They are good weeders, their feathers and down can be sold for extra cash, their meat is excellent (and abundant), and don't forget *pâté de foie gras!*

And, yes, they are very interesting. To me, they are the reincarnation of an especially "rammy" ram. People we know had geese that were especially playful and in fact enjoyed playing with their children. One day as friends were visiting and they were riding their snowmobile one of the geese jumped on the back for a ride. Naturally, their friend, upon seeing this, jumped off and away the goose rode, honking madly, across the field and into a ditch.

BREEDS

There are breeds of geese to suit your various tastes: meat, eggs, eggs and meat, show and pets. The table gives the major breeds of geese and their size, as recommended by the American Poultry Association in the Standard of Perfection.

TABLE 9: Breeds of Geese and Their Weights

	Male		*Female*	
Breed	Young	Adult	Young	Adult
African	16	20	14	18
Buff	16	18	14	16
Canada	10	12	8	10
Chinese	10	12	8	10
Egyptian	5	5½	4	4½
Emden	20	26	16	20
Pilgrim	12	14	10	13
Sebastopol	12	14	10	12
Toulouse	20	26	16	20

African: The African has an erect body that is carried high from the ground. Its head is light brown with a large black knob atop. It has large dark brown eyes and a black bill. The feathers are light ash brown on the underside of the body and on the neck and breast and a darker ash brown on the wings and neck. African geese are good layers and they grow rapidly to market weight, but they have more pinfeathers than most meat birds. They also tend to be very noisy.
Buff: These have poor meat and egg-laying qualities and are not worth considering for our purposes.
Canada (wild): These are not good meat or egg birds and thus are valuable only for pets or show. You need a permit to own one, they can only be kept confined in a wire pen or by clipping their wings. They are better off left in the wild.
Chinese (brown or white): This is a good multipurpose breed. Their graceful, swanlike appearance makes them popular for show or pets; their relatively high egg production (40 to 65 eggs per year) makes them desirable egg birds; and they are of medium weight, they grow rapidly, mature early and their meat is delicious. As with all poultry, the white variety is more desirable for meat purposes because of the absence of dark pinfeathers. Chinese geese also make very good "watch-dogs."

Chinese

Egyptian: Due to small size, worthwhile only for pets or show.
Emden: A pure white, erect goose. They are reasonably good layers (35 to 40 eggs per year) and are good setters. They are

also good meat birds as they grow rapidly and mature early.

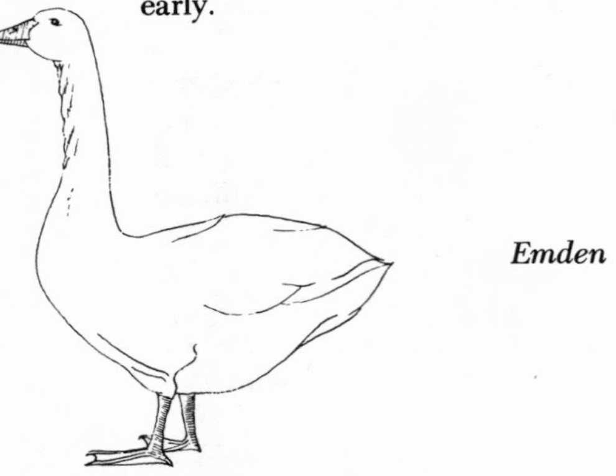

Emden

Pilgrim: The Pilgrims are good market birds and they have the added distinction of being sex-linked (color-coded by sex), a decided advantage for those of us who are not terribly proficient at fondling a goose's sex organs (goosing a goose?) to determine sex. The day-old male goslings are creamy white and the females gray. The mature ganders (male) are white with blue eyes, the geese (females) gray and white with dark hazel eyes.

Pilgrim

Sabastopol: This lovely white goose with curly feathers is good only for show or pets.

Toulouse: Toulouse geese make good mothers and lay about 25 to 30 eggs per year. They are also quite good as meat birds. They have broad bodies and look even larger because of the fluffiness of their feathers. They are dark gray on the back and the breast and white on the abdomen. Their eyes are dark brown to hazel, the legs and toes are deep reddish-orange and the bill orange.

Geese, too, can benefit from crossbreeding. A particularly good cross is a White Chinese gander with an Emden goose. The result is a rapidly growing market bird. Experiment.

PURCHASE

As with all poultry, you can buy fertile eggs for hatching, day-old goslings or mature stock for breeding. If you do plan to go into breeding, get sexed birds or buy enough so that you're sure you have a gander (preferably two, in case of accidental loss), and you won't have to buy one when your geese are grown. Flocks of geese tend to be very tight-knit so if you plan to introduce a new member prepare yourself for a battle.

Hatching

In an incubator the procedure is the same as for chickens, except that the hatching time for Canadian and Egyptian geese is 35 days and 29 to 31 for the other breeds. Incubators are difficult to use because more humidity is needed than for hatching chickens. As always, have a pan of water on the floor of the incubator, and you might try dipping each egg in lukewarm water for half a minute each day. Hatchability decreases after the first week, but eggs can be stored up to two weeks before incubation without losing viability.

A goose or a duck foster-mother hatcher (preferably a Muscovy) can handle 10 to 12 eggs. Broody hens must be free from lice, and because of the size of the eggs an average hen can only handle four to six. The eggs need not be sprinkled with water if the goose or duck doing the hatching has daily access to swimming water. In the case of chickens or ducks or geese without access to water, dip the eggs in water daily as in incubation. Eggs must be turned daily as they are too big for a hen to turn herself. Also, when a hen is used, goslings must be removed from her as they emerge and returned when all of them have hatched. If she deserts them or wanders away, confine her with them, as ducklings with a chicken surrogate.

Day-Old Goslings

The brooder can be the same as for chickens but allow ½ square foot per gosling at the start, and increase to 1 foot at the end of two weeks. Depending upon the weather, artificial brooding is only needed for a couple of days to two weeks. If the weather is warm and sunny, you can let the gosling's out after a few days, but be very sure they don't get caught in a shower or exposed to heavy dew. If you don't have a chicken brooder, you don't need to build one as elaborate as for chickens. Because they are so hardy, goslings can be brooded in a box with a light or in the corner of a room or barn. Make sure that no predators such as rats or weasels can get to them.

HOUSING

In most cases, housing for geese is minimal or none at all. Geese raised in warmer months need a simple shelter so that they can be confined at night safe from predators; if predators are not a concern, they may be left to find their own shelter. A more permanent shelter for breeding purposes can be the same as indicated for ducks. Allow at least eight square feet per goose for the larger breeds. Don't worry about them in winter as they love to frolic in the snow and may spend more time outside the pen than in it.

Don't just plop your geese down anywhere. Geese will roam quite a bit, so it is a good idea to house them as far from your garden as feasible. They can be quite messy with their droppings and graze very closely, so rotation of pastures (if fenced in) or a location out of barefooted humans' traffic patterns is recommended. If you are forced to fence your geese because of predators or angry neighbors, do so with a three-foot fence. Geese will also tear the bark from young trees, so keep them from young fruit trees (or other trees you don't want to lose) or wrap them with wire as high up as the geese can reach.

EQUIPMENT

Feeding and watering equipment is the same as for ducks. The range feeder is especially important for large groups of geese. The waterer should be a bit deeper so that they can immerse their entire head, but not so big that they can swim in it. Swimming troughs, like those for ducks, can be provided if there is no natural water. Remember, do not let them swim before three or four weeks of age or chilling may result.

Your nest boxes can be located inside or outside the house because more often than not geese like to lay their eggs outdoors. If the boxes are outside, do place them in some sort of natural shelter or build a rudimentary one to protect them from the elements. You will need one nest for every three birds, and it should measure two feet square. If possible, space the nests a distance apart from each other to discourage fighting.

FEED

Fresh water must be available at all times.

As with ducks, geese should not be fed medicated chick starter. Try to find a duck or goose starter (20 to 22 percent protein) and feed that for the first three weeks. If you can't find any starter, you can feed them bread crumbs wet with milk or a mixture of half cornmeal and half bran or oats moistened with milk. If the weather is suitable, you can let them out to forage for greens, or in the case of inclement weather supply them with greens in the brooder (along with their regular feed).

The rest of their feeding program is the same as for ducks, except that you can rely more heavily on feeding whole grains and they will forage a great deal more than ducks do. Mix whole grains with the mash or pelleted feed at a fifty-fifty ratio. If you wish, for greater economy, you can feed only whole grains, corn, oats, bran or wheat. If you are not feeding commercial feed with grit, you should supply some to your flock if they don't get enough in their foraging.

Geese favor the more succulent, tender grasses like clover, eschewing old pasture and hard stems and even alfalfa. Geese can get by on almost no grain at all the first month (and they don't seem to eat very much anyway), and therefore can be raised for very little cost. If you raise your geese on forage alone, finishing will be important for the best quality meat. If you confine your birds to a run and they exhaust their greens, it is essential that you supply them with plenty each day.

Finishing

A month or three weeks before slaughter gradually work your geese over to a diet of ½ to ¼ pasture and the remainder fattening feed. This finishing feed can consist of whole corn, cornmeal, barley, or a cornmeal and mash (or pellet) mixture moistened with milk. What you use depends mostly on what is available and cheap—experiment! As with any animal, do not change their feed overnight.

Breeding Diet

If you are wintering geese for breeding (or if you have become attached and want them as pets), you can feed them as for breeding ducks or a diet consisting of 15 to 25 percent breeder feed and the rest good, tender hay (preferably legume—second cutting is best) or silage.

Supplementing Commercial Feed

The foraging ability of geese makes them one of the most economical meat producers to raise. In addition, much of the feed in programs outlined above you can grow yourself. You could plant a field of clover to supply exceptional grazing for your flock. Allowing geese to forage and feeding them corn (or barley, oats, etc.) that you grow yourself should reduce feed cost to an absolute minimum: the cost of the seed itself. With very little effort you could grow (or glean from a farmer's field) all the grain you need for half a dozen geese.

MANAGEMENT

Routines

Much has been said about the weeding talents of geese. They are good for weeding strawberries up until the green fruit appears. Before the berries begin to ripen you'd best move your geese or you'll have no strawberries left. They can also wreak havoc with a vegetable garden. They will weed, to be sure, but also "bean," "corn," and "lettuce" while they're at it. If you wish more information on the subject, write for *Weeding with Geese,* available from the Department of Poultry Husbandry, University of Missouri, Columbia, Missouri 65201.

Handling

Geese will follow you most anywhere if you coax them with a little grain. If you want to pick them up, secure them as gently as possible by the base of the neck, reel them in, and hold the wings gently with the other arm. Grabbing them by the legs, as with chickens, can cause injury.

Predators

Geese are liable to the same predators as other poultry but are able to ward off the smaller and more timid animals. Our dog, chicken killer of repute, has yet to tangle with a goose. They may not be able to fight off large dogs, but they are convincing bluffers. Still, a good house to lock them in at night, and as a last resort, a pen, will do much to prevent losses.

Goslings are, of course, more prone to predator loss. Rats and weasels especially can wipe out a whole flock if they can get at them in the brooder. As with all poultry, make sure the brooder is predator-free.

BREEDING

Select geese that are healthy and rapid growers for mating. One gander will service about four geese. They are not necessarily monogamous as people say (except the Canada Goose, with which we are not really concerned), but once they do establish mating partners, it is difficult to introduce new geese to the flock. Generally, the laying and meat birds we are talking about tend to mate in pairs or trios (i.e., one gander to two or three geese), while the smaller breeds may mate one gander to four or five geese. The best practice is to have plenty of ganders and watch their breeding habits to see if any of the ganders are unneeded. If they are, butcher them.

Sexing, in all breeds except the Pilgrim, is not the easiest thing in the world. To determine sex (and this takes practice), take a bird and place it on its back (pointing the tail away from you) and bend the tail over backward a bit. Massage the genital area to relax the muscles and then press around the area with the thumb and index finger to expose the sex organs. The goose's will be rounder and more prominent while the gander's is less round. Practice on birds whose sex you know and in time it will be routine. This also works with ducks.

The Chinese may begin laying in the winter, but most geese will begin to lay in February or March as the days lengthen. They can be brought into production at the age of seven months using the same method of increasing the light as with ducks. If a good pelleted breeder feed (or mash) is available, begin feeding that a month before production. If no duck or goose breeder is available, chicken laying mash or pellets can be used. If you are feeding a home-prepared feed be sure to supply calcium (in the form of ground limestone, oyster shell, etc.) for eggshell hardness. Gathering eggs frequently will help cut down on broodiness. Hatchability will increase after the first year of production and is at its peak between two and five years. Geese will lay until about ten years of age (and will often live many years past this), and ganders are good breeders for about five years.

HEALTH

Geese are among the most disease-resistant of all livestock. Normal preventive measures should insure your flock against losses due to disease. The *Appendix* lists common poultry diseases, should you have problems.

BUTCHERING

Geese are ready for butchering at about 14 to 16 weeks. You can butcher at 12 weeks if you feed them extra heavily, but then again you're not saving money (and in fact are probably spending more), just time: and who's in a rush? Also at the earlier age you'll have to do battle with more pinfeathers, and that is certainly no delight. The best time to butcher is in the fall as they reach market weight and the supply of forage is dwindling.

Slaughtering is carried out as with other poultry. Save the blood (there's enough of it with geese to make it worthwhile) and the offal for other stock. The feathers are the worst to remove of all the poultry, but scalding and coating with wax then finishing the job up with a blowtorch should make an attractive carcass. Some people hang the gutted carcass for a few days to age; suit yourself.

Down

There may be some market for the down in your locality, or you can most certainly make use of it yourself for stuffing pillows, quilts, sleeping bags or your own parkas. If you wish to save the down, don't use wax when plucking. Wash the feathers in lukewarm water and detergent or water and borax and washing soda. Rinse them out and mat dry, then hang in an airy place in a cheesecloth bag.

You can also pluck live *mature* geese each spring and "harvest" the down year after year. Only mature geese should be used, and there will be no pinfeathers. Restrain the goose and pluck the down parallel to the body and toward the tail. Do this only after the weather is settled in the spring and do not pluck them bald. They get embarrassed. Wash and dry the feathers as you would after plucking dead geese.

IV. TURKEYS

"When God gave out brains . . ." the old saying goes. Well, that day the turkeys got tied up in traffic. Ditto when they gave out resistance to diseases. In modern parlance, a "turkey" is a worthless individual, an idiot, a dolt. These comparisons hold, however, only for domestic turkeys; wild turkeys seem to survive quite well on their own—out of the reach of man. It seems when we domesticate animals, "improve" them for our own purposes, we make them dumber. (Maybe we're the "turkeys" !)

Despite their apparent shortcomings, no one seems to find much fault with their taste. Take away an American's customary Thanksgiving or Christmas bird and he'll probably go into "cold turkey." Turkey is such a mainstay of American living that it was included in the first meal served on the moon. So much for the turkey that is commercially raised. But when it comes to the home-raised, freshly killed turkeys there is nothing—repeat, nothing—like their taste.

Turkeys are not the easiest poultry to raise, but they are definitely within the reach of the average homesteader. They require little space, little time, and with sanitary conditions they should be disease-free. Most anyone who raises animals should consider raising a few turkeys over the summer—a couple for their own use, and a couple for sale to pay for the ones they keep and even make a little profit. If ever your turkeys get you down, if their stupidity amazes you, and you wish to be rid of them on the spot, be humbled by the old saying: "The only thing stupider than a turkey is the one who raises them."

BREEDS

There are three breeds that you will be most likely to come in contact with: the Broad-Breasted Bronze, the Broad-Breasted White, and the Beltsville White. The Bronze is the breed everyone pictures when they think of turkeys. Their plumage is brown or black, and because of the problems in

Broad-breasted Whites

plucking a dark-feathered bird (getting a clean-looking carcass), the White has pretty well replaced the Bronze as a commercial bird. The Beltsvilles are smaller and make good broilers at 15 to 16 weeks and good medium-size roasters at 22 to 24 weeks of age. The larger birds (Whites and Bronzes) make good large roasters at 24 to 28 weeks. The large-type hens can be butchered at only 13 weeks as a fryer-roaster, and can be butchered at 20 weeks to make good medium-size roasters. The main difference between the Beltsvilles and the larger Whites and Bronzes is that the Beltsvilles do not convert feed quite so efficiently, but they are usually cheaper to buy as poults (young turkeys).

PURCHASE

You will be able to choose between the purchase of eggs and poults. Breeding turkeys is beyond the scope of small operations. Turkey eggs can be incubated much the same way chickens' eggs are, except that the hatching time is 28 days rather than 21. After hatching, or upon purchase of poults, you will have to place the turkeys in a brooder. Brooding is the same as for chickens. (Note: turkeys must never be housed with chickens, nor should they be kept in the pen or brooder that has housed chickens within the past three months or where chickens have ranged within three years.) These precautions are to prevent turkeys from contracting diseases. If you use a brooder or pen before the time has elapsed, disinfect it thoroughly.

Allow the poults twice as much space and headroom as you would for chickens. Turkeys are more sensitive to drafts and dampness, so take pains to eliminate such conditions. Also, cover the litter with rough paper for a week to prevent litter eating. Litter is not needed if you raise them off the floor on wire or wooden slats. Weather permitting, they may be removed from the brooder in eight weeks. Round corners in brooder pens with curved cardboard or hardware cloth, to prevent piling up in corners and smothering.

Purchase your eggs or poults from reliable breeders who guarantee disease-free stock. While you can order turkeys by mail, they do not ship as well as chickens and you should try to buy locally before you go the mail-order route.

HOUSING

There are three types of housing: confinement, sunporch, and ranging.

Confinement

This is very popular because it protects the birds from most predators as well as severe weather conditions. Again, dryness is one of the primary concerns. Litter must be deep and dry and more added as it becomes damp. The birds can also be raised off the floor on heavy hardware cloth or on boards that are slatted ¾ inch apart. Roosts are not necessary, but if they are desired construct them as you would for chickens. Allow at least one foot of space per bird on the roosts. In confinement housing allow five square feet of space per bird if they are debeaked, as described in the chapter on chickens, and seven to eight if they are not. The main drawback of this system and the sunporch method is that the turkeys won't be able to make use of greens and other supplementary feed the way range-reared birds can.

Sunporch Method

This is really identical to the above with the addition of a wire- or slat-floored "sunporch" attached to the house.

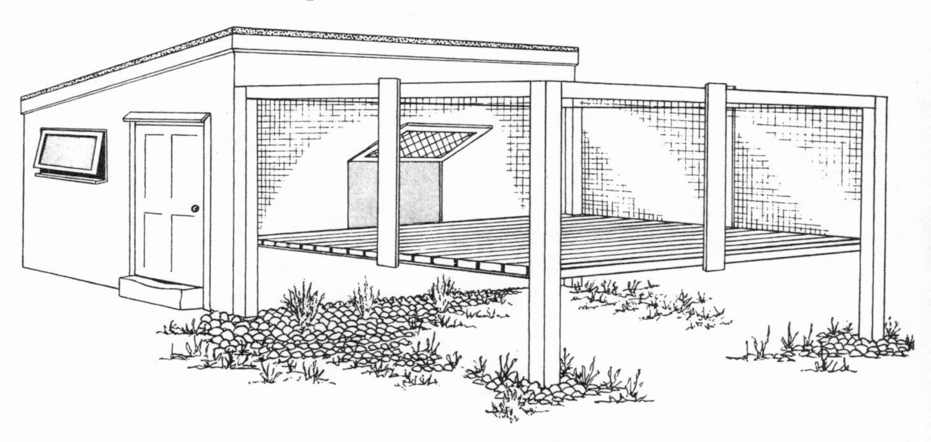

Ranging

This can be accomplished simply by letting the birds out of their confinement house and allowing them to free-range. If you have chickens they *must not* mix. You can keep them apart by keeping them at different corners of the property as they usually don't range too far from their houses. If they still get together, simply fence in a run for your turkeys. Avoid ranging them on poorly drained land as the wetness may contribute to disease problems. Lock up your flock at night so they are not threatened by bad weather or predators. In warm weather they can be outdoors at eight weeks of age; in more severe weather wait until 10 or 12 weeks or until weather has settled.

FEED

An average turkey will consume about 75 pounds of feed from birth to butchering, but you can be consoled by the fact that it has an excellent (4:1) food conversion ratio.

Poults

From birth to the age of eight weeks young turkeys need a 28 percent protein mash fed free choice. Mash must be fed up to four weeks; therafter a pelleted feed may be provided if you wish. Supply grit early and at all other times when a whole grain is fed and the birds are confined and don't have access to natural grit. Water should, of course, always be available. Young birds may need to be taught to eat. Tempting food such as coarser-milled chicken scratch or shiny marbles will usually attract them to the food and get them started.

Growing

From eight weeks on until slaughter the turkeys require a 20 to 22 percent protein growing ration. This comes in mash or pellet form and should be fed free choice along with some grain (corn is best, but anything that is palatable will be fine), also on a free-choice basis. Up to 16 weeks whole grains should be cracked, but can be fed whole thereafter. Greens can make up a large part of a turkey's diet. If you don't range your birds supply them with *fresh* greens once or twice a day. Free-ranging birds receiving fresh air, sunshine and plenty of greens will require a less complete feed (no vitamin supplements needed) than birds raised in confinement. If a lot of greens are fed to your turkeys, it is advisable to finish birds before slaughter. For two weeks prior to slaughter they need only a 16 percent protein ration. Corn is the best for finishing and should be supplied free choice.

Supplementing Commercial Feed

Greens, available to free-ranging birds or supplied to those in confinement, can supply up to 25 percent of nutritional needs of your turkeys. Good greens include Swiss chard, alfalfa, tender grasses, grain sprouts, rape, lettuce, cabbage, etc.

For the size flock most people will have it would hardly be worth the trouble to assemble your own mash as you might with chickens. Even if you did make your own chicken mash, it wouldn't be interchangeable because of the lower protein content. You can, however, make good savings by growing some of the grains fed during the growing period.

Surplus milk can be mixed with the mash, but be careful it is cleaned up quickly and does not sour. Although I have never tried it, you can try feeding turkeys high-quality food scraps (equal to Grade I foods; see chapter on *pigs*) much as you would chickens. Because they demand a higher protein ration than chickens or pigs, the scraps should be rich in high-protein food such as meats, cheeses and the like. Experiment. Do be sure that the feed is fresh and the birds don't sling it around the cages. Spoiled feed will invite predators and disease. It is a good idea to feed such scraps only to birds on range for reasons of cleanliness.

MANAGEMENT

Routines

Since colder weather may be a problem in raising turkeys, a spring to fall program is recommended. Buy your poults as soon as the weather is settled in the spring. Get a half dozen and figure on keeping some for yourself and selling the rest. You should have no trouble selling them and might even turn a bit of a profit. Incidentally, hens bring a slightly higher price than toms.

Handling

Much the same as with chickens. If you plan to "herd" them into a corner and grab one, be careful as such practices may cause them to panic, stampede and injure themselves.

Predators

Predators are much the same as for chickens. Again, a well-constructed, predator-free cage that you can enclose the flock in at night will be your best insurance.

BREEDING

Breeding turkeys is really beyond the scope of most homesteaders (and this book). If you do want to try it information is available in a good poultry book, such as Leonard Mercia's "Raising Poultry the Modern Way."

HEALTH

Strict sanitation and keeping your birds dry and free from chills will go a long way to keeping your flock disease-free. Remember, do not let the turkeys mix with chickens and thoroughly disinfect any equipment that has been used by chickens in the three months before using it for your turkeys.

Medicated turkey feeds (which includes most, if not all, commercial feeds), if you're not against using them, should be quite helpful in preventing the two most common diseases: Blackhead and Coccidiosis.

The Poultry Disease Chart (see *Appendix*) should be helpful in spotting and controlling most common diseases.

BUTCHERING

Butchering is the same as for chickens except that the oil sac on the back near the tail should be removed as it can cause an odd flavor.

CHAPTER TWO

Rabbits

Instead of taking your maturing Easter bunnies and hurling them out of the window on the way to work or allowing your children to maul them to death . . . eat them! Better yet, get a buck and a doe or two and begin raising your own. One ten-pound doe can produce in her litters up to 120 pounds of meat per year—a production of over 1,000 percent of her body weight! (To give you an idea of what this means, consider a good sow weaning two litters of ten 25-pound pigs per year. This would be a production of about 100 percent of her body weight. In order to equal a doe's output she would have to wean 200 such piglets or 5,000 pounds of piglets!)

Rabbits make an excellent backyard venture. They do not require a great deal of care, and they are efficient converters of feed to meat. The litters are butchered before weaning, so that there is no need to go the additional time and expense of building more hutches. For those of you who have labored over the job of plucking chickens (and inhaling stray feathers), butchering rabbits will be a welcome change. A litter of eight to ten rabbits can be dressed in an hour or two. The meat is similar, though I think superior, to chicken. It is all white and has a more delicate flavor. It exceeds the protein content of beef, pork, lamb and chicken, but has a lower percentage of fat, cholesterol, and ounce for ounce, fewer calories. The finished product is ready even sooner than veal, and because of the small bones there is less waste (only 20 percent) in butchering than any other animal we will cover. As the world food picture grows gloomier, the rabbit will emerge as the animal of the future.

May your rabbits be fruitful, and multiply. . . . .

BREEDS

There are over 66 breeds and variations of rabbits in this country, but we need only be familiar with a small number. Rabbits can range in mature weight from 2 to 3 pounds for a Polish to up to 20 pounds for a Flemish Giant. There are four weight classes of rabbits:

TABLE 1: Weight Classes of Rabbits and Common Breeds Within Each

Giants	*Medium Weight*	*Small*	*Dwarf*
Flemish	New Zealands	Tans	Polish
Checker Giant	Satin	Dutch	Netherland
Chinchilla Giant	Champagne D'Argent	English Spot	Himalayan
	Californians	Havana	
	Chinchilla		

While one might think the Giant breeds would be the logical choice for meat breeding, they are too heavy-boned and thick-skinned and too poor in feed conversion to be economical for meat use. The medium-weight breeds, as listed above, are light-boned and thin-skinned so less food goes into the production of skin and bones and more into meat. Most breeds within this class produce rabbit-fryers of three to four pounds in eight weeks. The Small and Dwarf rabbits are most often used for show and laboratory purposes.

New Zealand White

Californian

The hides of the medium-weight breeds, although not as high quality as the fur breeds, are nonetheless suitable for homestead use. Your choice of medium-weight breeds will probably be limited because in most parts of the country rabbits have yet to be accepted on a large scale as meat producers. By far and away the most popular and available rabbit is the New Zealand White. They are hardy and fast gainers and therefore the most desirable.

PURCHASE

If, after you decide on a breed, you can't locate a source in your locality contact your county extension agent or write the American Rabbit Breeders Association (see *Appendix*) for sources of your choice in your area. You might also check agricultural market bulletins or look for rabbit clubs. If you still cannot locate a breeder (which is unlikely), you might check in the back of the rabbit publications (listed in *Appendix*) for mail-order services. This is more expensive and your chances of loss are higher, but it will give you a wider selection of breeds.

Age

Since rabbits don't reach maturity until five or six months of age, if you buy weaned rabbits (eight weeks old or so), you'll have a bit of a wait before you can begin breeding. On the plus side, they are cheaper than adult rabbits and by buying young stock you will ease into the rabbit business, giving yourself extra time to iron out any problems before you have to worry about breeding. If you are buying young stock, try to do so as early in the spring as the weather will permit so they will be old enough to breed in the summer. Cold weather breeding can be a problem, especially for first litters.

If you are ready to get right into breeding try to buy some five or six-month-old stock, but be prepared to pay a slightly higher price. Age is rather difficult to determine accurately in rabbits so you'll have to rely on the word of the seller (or try to get a look at a particular rabbit's hutch card if there is one). You won't run the risk of an overage animal if you buy from a friend or reliable dealer. Older rabbits should be avoided, because if an older doe has not been mated for some time she will often go sterile. (Again, if the hutch card is available, check it.) Your older bargain basement doe may never produce a litter. Old rabbits can often be spotted by their long and heavy toenails. Also, if the meat along the backbone is sinewy and tough as compared to younger stock steer clear.

A good size herd to begin with is one buck and two or more does. If you plan to have more than ten does then you will need two bucks as one buck is able to service up to ten does.

Sexing Determining the sex of a young rabbit is impossible with the naked eye, and the correct procedure is difficult for the novice. To sex a rabbit, place it on its back and use the index finger of one hand to hold the tail out of the way. Using the thumb and index finger of the other hand, firmly but gently press down and manipulate the sex organ so that you expose the reddish mucous membrane. In bucks the membrane will protrude to form a circle, while in does it will form a slit with a small depression toward the rear. An easier, although less reliable method of sexing is to hold the rabbit by the scruff of the neck and feel along the belly. If the area is smooth it is a

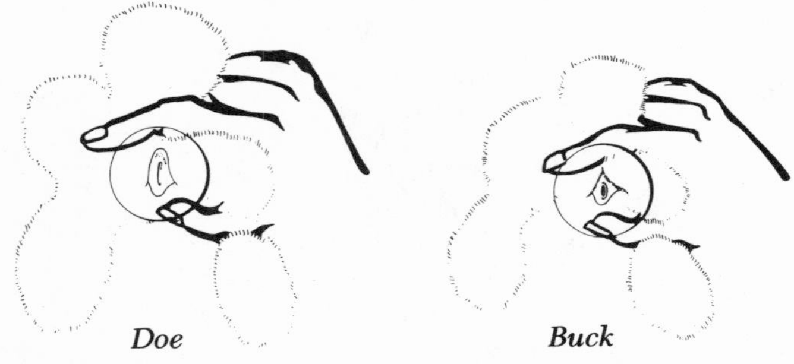
Doe *Buck*

doe, if it's bumpy it's a buck. In older rabbits, the testicles will be prominent in males so sexing presents no problem. With enough practice you will be able to sex a rabbit quite easily. In the manipulative technique, do not use any one rabbit for too long a time as irritation may result. Put yourself in his shoes.

In selecting breeding rabbits for meat, look for ones that are chunky and blocky as opposed to the long, rangy ones. They should have large feet without any sores on them. Similarly, the ears should be free from sores and the mouth should be well formed and not have teeth that are broken or "buck teeth" (malocclusion). The rabbit should appear alert, have bright eyes and a smooth, glossy coat. Inquire about the rabbit's parentage as their characteristics were passed on to the rabbit and will in turn be passed on to its offspring. All large rabbitries will have extensive hutch records that you can look at to help you in your choice. (This is another good reason to buy from an established rabbitry.) Look for stock that comes from large litters whose mother consistently weaned from seven to ten young. Few deaths prior to weaning would also indicate a good milk supply as well as disease resistance. Studying a hutch card will also give speed of growth as well as the number of litters per year and other pertinent data. Remember, as with other stock, it is better to settle for the worst from a good litter than the best from a poor one.

HOUSING

Location

Your rabbit pens should be located in a place free from drafts and protected from rain or snow. Rabbits can stand a great deal of cold but cannot survive windy, drafty conditions. Because they do not need a great deal of sunlight, and in fact should not have very much, they can be raised in the semidarkness of a barn, garage or any available space in an outbuilding. In warmer climates their pens need only have a watertight roof to keep out rain. If you keep them outside, take care not to have their pens exposed to too much sunlight. In very warm weather you may have to cool their pens by draping the sides with burlap soaked in water.

The best sub-hutch flooring is dirt or concrete. Wood is the least desirable because it absorbs urine and doesn't allow proper drainage. Concrete is adequate but necessitates more frequent cleaning and is relatively expensive if you have to install it. Dirt is good because it allows proper drainage and needs a minimum of maintenance. Wood chips or sawdust sprinkled under the cages will tend to absorb urine and reduce odor. Peat moss is also a good material to place under the cages as it has high absorption properties and it is also a good medium in which to grow worms for possible sale (see *Management*).

Pens

Whether your pens are of the all-wire Quonset-type, metal and wire construction (see *A* and *B Figure 3*), a design of your own choosing, or a commercially purchased hutch (see *Appendix*), it will need to be located correctly as indicated above. Allow one square foot of space for each pound of live weight of the rabbit(s). I happen to favor the wire/wood design as illustrated because it can be easily moved and even be placed out of the barn in suitable weather. I also like the hay manger in the center, as hay makes up an important part of a feeding ration (see *Feed*). You can make a manger in an all-wire hutch by making a holder with chicken wire and hanging it from the ceiling. Remember to make your cage larger if you plan to hang a manger. If you use feed dishes that sit on the floor, allow for them when figuring your space requirements.

Some people report foot problems (sore hocks) with the use of wire-bottomed cages. If that happens you might try spreading some straw on the wire or place a piece of board in the cage for the rabbit to stand on. Also, in the construction of the cage, note that the wire has hardened zinc drippings from when it was dipped in the process of galvanizing. Install the wire "drip side" down or these drippings will stick up and act like tiny spikes and injure the feet. If sore hocks persist, you can put in a wood floor that slopes to the rear of the cage leaving a 6-inch-wide section of ½-inch hardware wire. Most of the urine and feces will run off the wood and out the floor through the wire. In cages with wood floors, more care will have to be taken to keep the cage clean.

Another cage design I've seen that might be useful for temporary use (as in the case of weanlings you haven't quite gotten around to slaughtering yet), or for a buck or dry doe, is a "bottomless" cage. It's simply a wooden frame about three feet by four feet with chicken wire on the four sides and the top, but with the bottom left open. This can be moved around inside on the barn floor or even placed outside on the ground

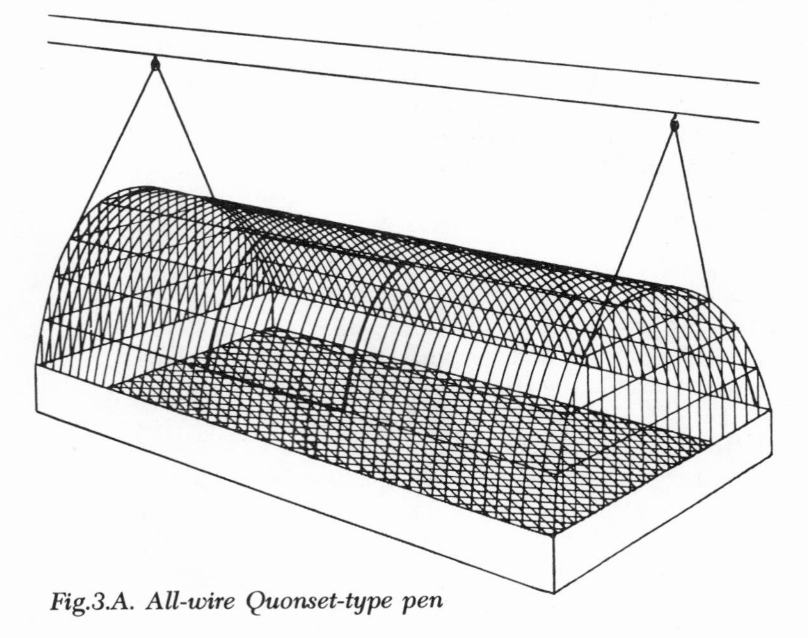

Fig.3.A. All-wire Quonset-type pen

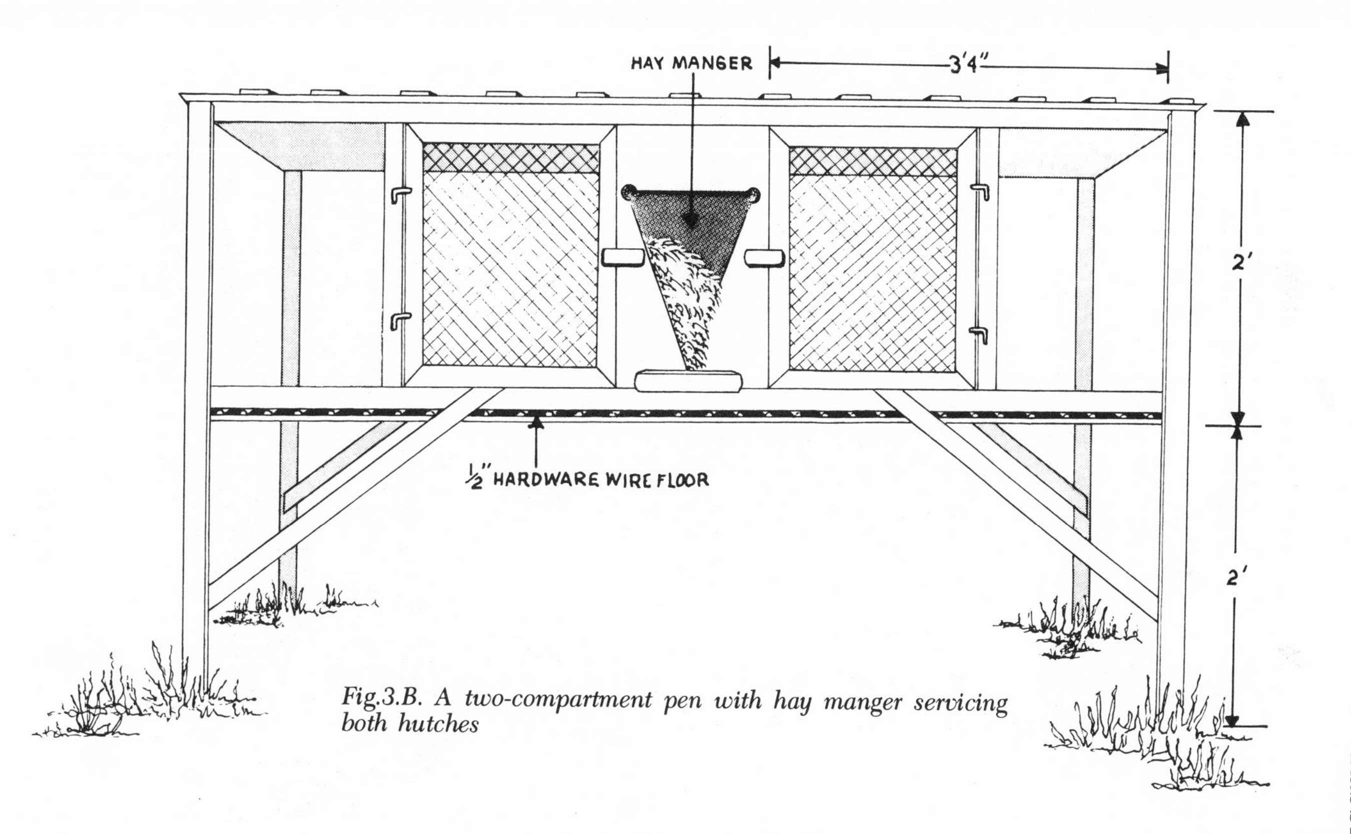

Fig.3.B. A two-compartment pen with hay manger servicing both hutches

in warm weather. To feed or water simply lift it up and slip the dishes in. While this really isn't suitable for permanent use, it is convenient if you need an extra cage that is portable and very cheap.

Remember in your cage construction that rabbits will chew any exposed wood. Also, wire spread over wood on floors will tend to trap urine and feces. You will have to clean these areas often with a wire brush or avoid such construction altogether. Tin or aluminum nailed over exposed wood will reduce chewing.

EQUIPMENT

You can buy self-feeding hoppers, ceramic water dishes or automatic waterers and nest boxes from suppliers, and there is nothing wrong with them except they will undoubtedly run into money which will jack up the price of your meat. Many homemade substitutes can be constructed that will work just as well.

Feed Dishes

The biggest problem with feed dishes is that they tip over. Feed is wasted and contaminated by rabbit droppings. You can simply use a coffee can or other type of can, nailing it to a board or wiring it onto the outside of the cage to prevent tipping. Self-feeders that hold more than a day's supply of food are more costly, and because exposure to air and humidity makes the food soft and unpalatable and also robs it of nutrients. If you fill the receptacle with a day's feed allowance, exposure to air shouldn't be a problem. On the other hand, what's the use of a self-feeding hopper that must be filled every day like a simple dish? If you do construct one, which is advantageous if you have a large herd of rabbits, have a hopper that spills feed into a trough below. Having the hopper on the outside frees you from having to reach in the cage when filling. It's a good idea to place a piece of window screen about a half inch or inch off the floor of the trough so that dust from the feed will fall through and not interfere with the rabbit's breathing and eating.

You might also place a removable trough that is temporarily secured to the floor (or your rabbit friend will knock it about the cage), under the hay rack to catch any hay that falls from the rack. This will prevent undue wastage and you can also provide your pelleted feed in this if you wish.

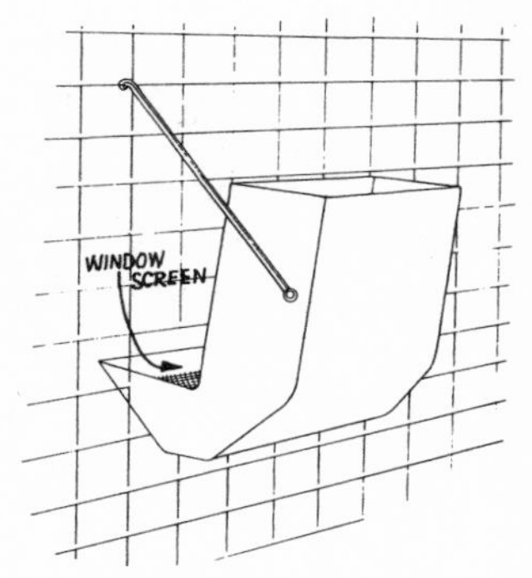

A self-feeding hopper attached to a hutch

Those heavy ceramic feed dishes you can buy in feed stores are ideal for rabbits. Their weight prevents tipping but the price will preclude their purchase if you have a lot of rabbits. You can make a type of concrete dish that, while not quite so durable as ceramic, still has its benefits at almost no cost to you.

A small bag of ready-mix cement, needing only the addition of water, a plastic dish (the kind coleslaw or dessert topping comes in), and a small amount of lumber are what you will need. Make a wooden form, held together with screws, an inch wider than the dish. Grease the sides of the form with motor oil before each use. Rest the dish on small stone or block of wood—at least an inch above the bottom of the form. Pour the cement in the space between the dish and the form. When the

1"-2" BETWEEN DISH AND SIDE OF FORM

A reusable wooden form and plastic feed dish set in concrete

cement is dry lift off the form. If your homemade dishes crack, add a few pebbles to the next batch and reinforce with scraps of hardware cloth in unset cement, making sure no ends protrude that might injure your rabbits. If the rabbits chew at the plastic inner dish or it cracks, remove it and use the concrete dish that is left.

Creep Feeding

Some rabbit raisers are in favor of creep-feeding young rabbits so they will have attained greater weights by weaning time. While they grow quite adequately to butchering weight on the doe's milk alone, you may want to give them a boost with dry food. (Some companies—Carnation for one—makes a creep feed especially formulated for young rabbits.) A creep feeder will allow the bunnies to eat but will reject the doe and can be built quite simply and cheaply. It can be made of wood and should measure about eight inches long, four inches high and three inches deep.

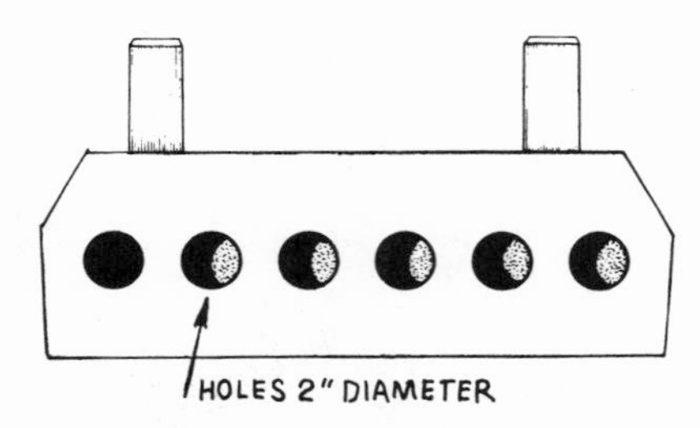

Rabbit creep feeder

There should be holes cut in the face two inches in diameter for their heads. If they chew the wood, the exposed edges can be faced with aluminum.

Hay Racks (Explained in the section on *Housing*.)

Watering Equipment

While automatic watering systems are the norm for large rabbitries, they are both impractical and too expensive for a small-scale project. Water is very important for rabbits, and in warm weather a doe and her litter will consume up to a gallon of water in a 24-hour period. Waterers should be large enough to hold at least half the day's needs (for your sake) and also be heavy or well secured to prevent tipping. A variation of the concrete-encased plastic dish using a larger plastic container is both satisfactory and cheap. For winter use it will help if the dish is concave on the bottom and has sloping sides so that ice will be pushed up as it forms and will not break the dish.

Self-watering units are a time-saver but to prevent stagnation should not hold more than a day's supply of water. A one-quart bottle wired to the outside of the cage and turned upside down to fill a can is adequate, but since it holds only about 1/4 of the possible consumption it must be checked periodically during the day. (see *Fig. 8*.)

A larger plastic bottle, such as the type used to hold bleach or milk, can be secured to the outside of the cage and allowed to fill a can. Metal or plastic tubing can be fitted tightly to the top of such a jug and allowed to protrude into the pen. The rabbit will soon learn that sucking will provide it with water. Dewdrop valves, the type used in automatic watering systems (available from feed stores or in the back of rabbit magazines), can be fitted into a jug or bottle and used for watering.

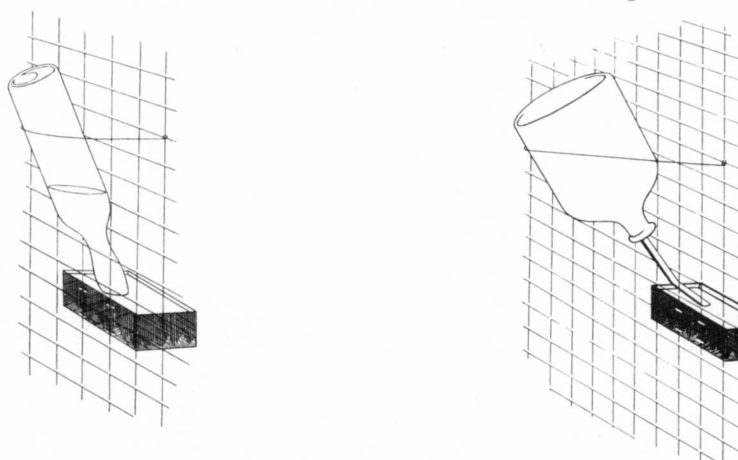

Fig.8. Two homemade waterers

To prevent water from freezing in the winter, hang a light bulb over a water dish. *Note:* Frequent cleaning of feed and water dishes is essential for the health of your rabbits.

Nest Boxes

Unless you have a very strange buck, nest boxes will only be needed for does, and they should be placed in the pen five to seven days before they are due to kindle. Almost any wooden or metal box that measures one foot wide, two feet long and one foot deep will do. The type of nest box I prefer is illustrated in *Figure 9.*

It affords good protection for the young, without being too stuffy or allowing condensation to build up. I recommend boring a few ½-inch holes at the rear of the top for further ventilation. The doe might also rest on top of the box, which many seem to enjoy.

One drawback for small operations in colder climates is that rabbits are rarely (if ever) housed in heated buildings. Since the young cannot be successfully kindled in nest-box temperatures below freezing, you might be limited to three litters a year instead of the desired five. This cuts down quite a bit on

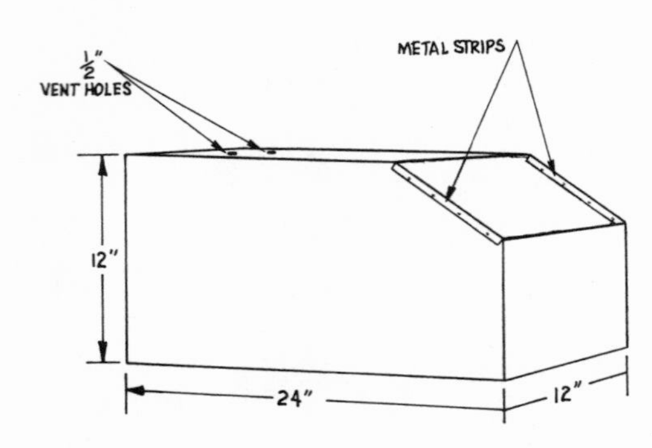

Fig.9. Nest box

the meat you can expect from your doe, and since you have to feed her in the winter anyway, will raise the price of producing your meat. You might try an insulated nest box, said to be used successfully, if the pen is draft-free, in climates where temperatures fall as low as 20 below zero. Construct a box as illustrated above, cover the outside surfaces, even the bottom, with one-inch Styrofoam and then cover with wood or Masonite (to prevent chewing). You can use fiberglass insulation, but it is much harder to work with. Three holes should be bored at the top rear for ventilation. Before kindling, fill the box with bedding so that the doe has to burrow in (not packed too full or she'll begin tearing it out) to kindle. If this doesn't work, experiment with an insulated nest box fitted with one or two Christmas tree light bulbs for added heat.

Miscellaneous Equipment

A salt spool, available at feed stores, should be furnished in each cage. Also, a small block of wood thrown into the cage for the rabbit to chew on may prevent chewing on exposed wood. It also helps to keep their teeth in proper condition.

Last but not least, the hutch card. Since it is strongly recommended that you keep records for *all* your livestock, this should not come as a surprise. The card, located either on the hutch or in a record book in your home, should list such information as name of animal, date bred and by whom, kindling date, number born, number weaned, weight and other data. On a buck's card include all pertinent information on breeding and his offspring. These records will enable you to select good breeding stock and cull poor producers. (Such cards are often supplied by feed stores.)

FEED

Commercial rabbit pellets are higher in protein (18 to 20 percent) than most other feeds and are, accordingly, more expensive. However, the rabbits' good feed to meat conversion ability (3.5:1) makes it possible to raise meat economically using only commercial feed.

There is as much danger in overfeeding stock as there is in underfeeding. It is rather difficult to set any hard and fast rules for feeding rabbits, since their requirements vary with breed, size, age, weather conditions, etc. Generally speaking, a mature rabbit that is being normally maintained (i.e., is healthy and not pregnant) will require 3.8 percent of its body weight in feed daily. Therefore a ten-pound doe will eat about six ounces of feed per day (10 x .038, or .38 pounds). What does this really mean? Well, because of all the variables mentioned above, it should simply serve as a rough guideline for beginners so that they don't give a rabbit ten pounds of food a day or expect it to subsist on three pellets. As you learn more about rabbits in general, and your stock in particular, you will be able to tell if they're getting the correct amount. If possible, have an experienced breeder show you a properly maintained animal. Get the feel of it, especially around the ribs and backbone. If a rabbit person is not available for advice, pick out one of your own rabbits that is producing well and study it. If the flesh over the backbone and ribs is spare, increase the feed. If there is too much meat, reduce the feed and continue to experiment until you reach the optimal level. In time, with daily handling of your herd, you will be able to maintain your rabbits well and to spot quickly and correct any problems.

Feeding Does/Creep Feeding

In feeding pelleted feed to pregnant does and does with litters, allow them all they will clean up between feedings. Creep feeding for young rabbits prior to weaning is definitely not necessary, but is gaining some favor among rabbit raisers as a method of quicker and more substantial gains. The doe's milk is the perfect food for young rabbits and is all that is needed until slaughter. Its high protein content (15 percent) makes it one of the richest milks of all livestock. If you want to experiment with creep feeding, supply pellets free choice from a creep feeder. At least one company, Carnation, makes a creep feed especially for young rabbits that is higher in protein (22 percent) than most rabbit feeds. Again, as with all animals, experiment, comparing weights of those litters that are creep-fed and those that are not, and decide whether the added cost of feed was worth it.

You can feed your rabbits as many times a day as you want, but one feeding is adequate (as long as it's enough). *Be consistent.* Rabbits eat the most in the evening and night, so if only one feeding is supplied do so in the evening.

Contrary to popular belief (reinforced by cartoons of Bugs Bunny subsisting only on carrots), rabbits cannot live solely on carrots and greens. If you feed greens (see below), allow your rabbits only what they will clean up in 10 to 15 minutes in the evening.

Water/Salt

As mentioned earlier, a mother and her litter may consume up to one gallon of water in a 24-hour period. Fresh water should *always* be available. Salt, in the form of a salt spool, should also be provided free choice.

Supplementing Commercial Feed

In conventional feeding (i.e., the sole use of commercial feed) you have a complete food and won't have to worry about other nutritional needs. In supplementary feeding you must have some knowledge of rabbits' nutritional requirements to be certain that they are met. In simplest terms, does (pregnant or with litters) need a 20 percent protein ration. A pelleted feed supplies this or it can be provided in the following formula:

TABLE 2: Feed Requirements for Does

	% of ration (by weight)
"Protein"	20
"Grain"	39.5
Roughage	40
Salt	0.5
	100.

Dry does and bucks need about an eight percent protein ration. They can, of course, be fed commercial feed but they are getting much more protein than they need (20%). It won't hurt them—only your pocketbook. You can feed other things, as we shall see, but the general formula is:

TABLE 3: Feed Requirements for Dry Does and Bucks

	% of ration
"Protein"	8
"Grain"	31.5
Roughage	60
Salt	0.5
	100.

Commodities that make up each of the above groups are given below. Some of these you might be able to grow, or come by cheaply, still others you will have to purchase commercially. In any event, making use of the above formulas and experimenting with combinations will enable you to come by a food mix that is the best for your stock and is the most economical for you. Sources:

(1) *"Protein"* In order of palatability and nutritional value for rabbits are: peanut meal, soybean meal, linseed oil meal, hempseed meal and cottonseed meal. Whole soybeans, while high in protein, are often less palatable for rabbits. Pellet meals or pea-sized cakes of the above meals are best.

(2) *"Grains"* Cereal grains in order of desirability are: oats, wheat and grain sorghums, barley, corn and milo. Barley and oats should be rolled to prevent undue wastage. Old bread or other wheat products can be used.

(3) *Roughage* Any good legume hay (second cutting is best). It is eaten most efficiently if cut into three- or four-inch pieces. Carbonaceous hays such as Sudan grass, timothy and the like can be fed but the protein content of the feed should be increased a bit.

(4) *Salt* Salt can be mixed in with the feed at the indicated rates. Better, simply supply a mineralized salt spool for all your rabbits.

The Walters family of Campbellsport, Wisconsin, have developed the following feed program and reported good success with it. It may not perfectly suit your rabbits, but can at least be used as a starting point in formulating a more self-sufficient ration for your herd.

TABLE 4: Walters Feed Program

Summer

Dry Does and Bucks:

Morning and Evening—Greens and weeds from garden (see *TABLE 5* on Desirable and Undesirable Greens) No more than they will clean up in 10 to 15 minutes.

Evening—Grain supplement (4-6 oz. of raw grain consisting of 3-5 oz. oats and 1-2 oz. soybeans)

Pregnant Does and Does with Litters:

Morning and Evening—Weeds and grains (see above)

Evening—2 to 2½ cups of grain mixture consisting of 30% corn (omit in warmer months), 60% oats and 10% soybeans.

Fall and Winter

Hay and a grain mixture of 45% corn, 45% oats, 10% soybeans. Feed free choice to active does and enough to maintain dry does and bucks.

Sally Cook of Rochester, New Hampshire, reports success with raising rabbits on a program that includes millet, squash and apples. She grows a plot of millet that she feeds to her herd unthreshed and unwinnowed, allowing them to eat the hay along with the grain (which the rabbits husk themselves). She collects culled or damaged squash from her garden and from a nearby squash farm for winter feeding. She allows her rabbits as much of it as they will clean up in a day. They will eat the pulp but enjoy digging for the high-protein seeds inside. Apples that are picked in abandoned orchards and are left over from her cider making are fed during the winter. While she does not gain top efficiency and five or six litters per year, she maintains that her herd is healthy and provides her with good meat at a minimum of cost.

Other rabbit feed recipes that can be made up by the industrious homesteader, as supplied by the U.S. Department of Agriculture, are as follows:

TABLE 5: For Pregnant Does with Litters:

(1)	Whole oats or wheat	15%
	Whole barley, milo or other grain sorghum	15
	Soybean or peanut meal pellets or pea-sized cakes (38-43% protein)	20

Alfalfa, clover or pea hay	49.5
Salt	0.5
	100.

(2)	Whole barley or oats	35%
	Soybean or peanut meal pellets or cakes	15
	Alfalfa or clover hay	49.5
	Salt	0.5
		100.

(3)	Whole oats	45%
	Linseed meal pellets or cakes (38-43% prot.)	25
	Carbonaceous hay	29.5
	Salt	0.5
		100.

For Dry Does, Bucks and Young Rabbits:

(1)	Whole oats or wheat	15%
	Barley, milo or other grain sorghum	15
	Legume hay	69.5
	Salt	0.5
		100.

(2)	Whole barley or oats	35%
	Legume hay	64.5
	Salt	0.5
		100.

(3)	Whole oats	45%
	Soybean, peanut or linseed pellets or pea-sized cakes	15
	Carbonaceous hay	39.5
	Salt	.5
		100.

To make a feed mix assemble all the ingredients in the correct proportions and feed it to your rabbits. If they get picky, and eat only "favorite" parts, you will have to have the ration ground. Whenever you feed any mash to rabbits be sure to wet it down a bit as dust will interfere with a rabbit's breathing. Since they eat very little compared to pigs, any food surpluses you can come by will result in substantial savings. If you can grow corn and have access to stale bread you have two food groups accounted for (using the food program on page 68) and need only buy a protein supplement to complete your ration. Often farmers will, for the asking, allow you to glean their fields after harvest in the fall. People I know collect left-behind ears and for a few days' work have all the corn they need for a winter. To get corn off the cob, you will need a corn sheller that costs less than $10 and is available from feed stores or livestock supply catalogs (see *Appendix*). Oats need not be rolled as rabbits will shell them themselves, and for those rabbits that will eat it the chaff can make good roughage. There may be more waste when feeding whole grains, so it's a good idea to place a board under the feed dish to collect dropped food.

Vitamin supplements to supplementary feed are generally unnecessary because the vitamins A, D, E and B should occur naturally in all feeding routines. Vitamin A is present in some root crops (carrots, Jerusalem artichokes) and good hays. Vitamin D is present in hay that is sun-dried and vitamin E is present in dry hay, grains and protein supplements. Vitamin B is synthesized in the stomach of the rabbit, in a type of pseudo-rumination, and the vitamin is obtained by eating a small amount of mucous-coated feces that is passed at night.

As stated earlier, when fed with care, greens can supplement feed. Young rabbits should never be fed greens as they may fill up on them and ignore more nutritious food. In a limited way, greens are quite healthful, but overingestion must be watched because of their high water content. The following table lists good and bad greens:

TABLE 6: **Desirable and Undesirable Green Feeds and Root Crops for Rabbits**

Desirable	
Rapidly growing cereal grains	Green hay crops
Rape	Kale
Trimmings from leafy garden vegetables	Lawn grasses
Dandelions	Tender twigs
Sweet potatoes and vines	Filaree, afilaria or Stork's bill
Plantains	Malva or mallow, cheeseweed
Mangels	Carrots and tops
Swedes	Turnips
Kohlrabi	Sugar beets
Culled potatoes and peelings	
Jerusalem artichokes	

Undesirable

Burdocks	Miner's lettuce
Castor beans	Nightshade
Fireweed	Oleander
Goldenrod	Poppies
Horehound	Sweet clover
Lupine	Tarweed
Milkweed	Rhubarb

(Source: From *Domestic Rabbit Production* by George S. Templeton. Danville, Ill.: The Interstate Printers & Publishers, Inc., 1968, p. 59.)

MANAGEMENT

Routines

While not too many Americans are familiar with rabbit meat, once they have tasted it they usually become converts. Invite some friends over for a rabbit dinner, and I'll bet they will become a ready market for your surplus meat rabbits (or for some weanlings as breeding stock) and get started themselves. By selling your surplus meat you may be able to pay for your own meat and even turn a little profit. As an added plus, rabbit manure is one of the most valuable of manures. You can use it for your own flowers or vegetables or find a market for it with earthworm farmers . . . or maybe begin growing your own earthworms (see *Appendix*). The pelts, while not valuable commercially, can be put to many uses around the farm: gloves, hats, blankets, etc. *Do not* be tempted to sell little bunnies as Easter gifts for children. While it may be seen as a source of added cash, more often than not it also ends up in death or serious mauling for the rabbit. Rabbits as Easter gifts probably are the most ill-treated of all pets.

Chewing: To prevent chewing, and the consequent destruction of the wooden portions of your hutches, cover exposed wood with tin, wire or soak with creosote at the time that they're being built. As mentioned earlier, blocks of wood and twigs will help to satisfy the rabbits' urge to chew and also condition their teeth at the same time.

Molting: Chickens lose their feathers, rabbits their fur. This shedding and restoring of the coat normally takes place once a year in mature rabbits and in the first several months of age in young rabbits. It is evidenced (not surprisingly) by the appearance of large amounts of loose hair in the cage and on the wire. It may be caused, unnaturally, by disease, sudden high temperatures, rapid changes in diet or other traumas. Normally, molting should be a minimal concern except that extra feed will probably be needed and conception in does and production of sperm in bucks may be adversely affected.

Handling

The old picture one has of carrying rabbits around by their ears is a dangerous one as it will result in injured animals. It should be left to magicians who can presumably make their injured animals disappear. The correct way to pick up and carry a rabbit is by grabbing the skin of one shoulder in one hand and placing the other hand under the rump for support. The feet should always be pointed away from the body to prevent scratching.

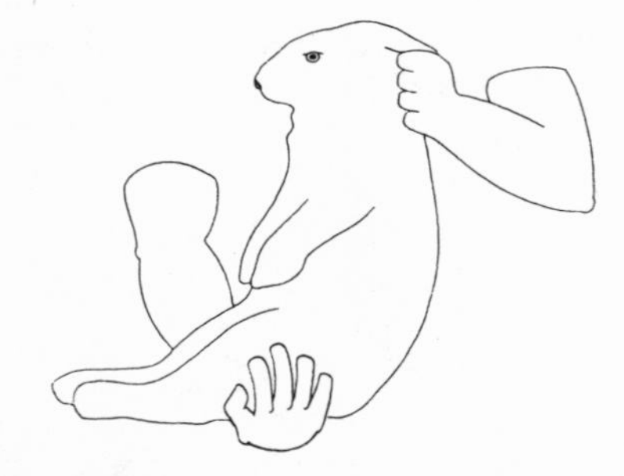

Predators

Cats, dogs, opossums, weasels, snakes, rats, etc., share with humans a love of rabbit meat. Housing your rabbits in a barn or other closed building will eliminate the most common predator, dogs. Well-constructed cages using wire should keep out most predators. Safeguard your young rabbits by having a thick bottom on your nest boxes to prevent rats and other animals from gnawing through and getting at them. If rats or snakes climb up the legs of your pen and into the cages, you should put circular sheets of metal around each leg to block their path.

BREEDING

The age of maturity varies with the weight class of the rabbit. The medium-weight rabbits, with which we are most concerned, usually reach sexual maturity by six months but sometimes as late as seven. The Dwarf breeds mature at five months and in the Giant breeds maturity is reached from the tenth to twelfth month.

The doe's egg cells are contained within small follicles. In nine to ten hours after sexual stimulation by the buck the follicle will break. The sperm that was deposited at the time of mating will fertilize egg cells within two to four hours after the follicle breaks.

The doe should be brought to the buck's cage for mating. This will tend to prevent fighting, as a doe is apt to attack a buck when he is brought to her cage. He will usually mount her without incident and, as with humans, when the service is completed he will often grunt and roll over. He will not, however, light up a cigarette. There is some controversy as to whether or not conception rates are improved by returning the doe for remating in a few hours. Some people say returning her will not improve conception but will only serve to deplete the buck. Others claim that since the sperm must be retained in the body until the egg follicle ruptures, the sperm may pass out of the body if the doe urinates. They think remating in a few hours (some also remate twice the next day) will insure pregnancy. Experiment.

In some cases the doe will not accept the buck. If she ignores him or growls do not leave them alone—they might fight and injure each other. Watch them for a while. If the doe evades him but her tail is twitching, she is probably playing hard to get and will accept him in time. If, however, the doe does not accept him, you might try moving her cage next to his for a few days so they can get used to each other. Often this will do the trick and she will accept him when he is reintroduced. Another method which often succeeds is to switch cages. After a few days bring the doe back to her cage, leaving the buck in. If you do not wish to resort to this mumbo-jumbo and want results *immediately*, you can try force mating. To accomplish this grasp the doe by the ears and the skin of the shoulder with one hand and place the other underneath her with two fingers on either side of the vulva, pushing it out a bit. As the buck mounts her, raise the hindquarters a bit to allow normal coupling. This may not always insure conception as there are four days in the doe's 16-day cycle when no eggs are available for conception. During these four days she also lacks interest in the buck and this may be why she rejected him in the first place.

The gestation period is an average of 31 days, but as short as 29 and as long as 35. Three to five days before kindling place the nest box and bedding (straw, hay, etc., but not sawdust as it can asphyxiate the young) in the hutch. Kindling is imminent when she begins plucking her fur to line the nest box. The first day after kindling offer her a tempting morsel of food and quietly remove the nest box and inspect the young. Remove any dead or deformed babies and any runts in a large litter. At this time you may even out litters by taking some babies from a large litter and letting a doe with a small litter nurse them. This can be done with litters which have kindled up to three days apart and is a good reason why you should breed your does to kindle at about the same time. While litters may range in size from 1 to 20, 8 to 10 is the ideal number for a doe to handle.

At about ten days of age their eyes will open and soon the babies will be in and out of the nest box. They will begin eating some of the doe's food and if you want, they can be creep-fed. At eight to ten weeks of age they should be weaned and slaughtered or some retained for breeding purposes. The nest box can be removed at four weeks, or may be retained longer in colder weather.

Problems

You may have a doe that will kindle her young on the floor outside of the nest box, or will refuse to nurse her young, or will kill them outright. Usually the cause is simple, such as moving her to a new cage too soon before kindling, excessive noise, or the presence of predators. Does are often quite sensitive close to kindling time and so it is wise to leave them alone as much as possible. If you do find young on the floor of the cage, don't panic. When I was young, one doe of mine kindled eight on the floor of the cage, but upon closer inspection I found that there were also ten in the nest box! I put them back in the nest box and she just as promptly returned them to the floor of the cage. Apparently, as is so often the case with livestock, it was a case of not giving her credit for doing what nature told her. She was evening off the litter herself, as she apparently knew if she tried to nurse all 18 they would die or at best be so terribly weak that they would fall victim to disease.

If she won't nurse and you find there is no milk, mastitis may be the problem. This occurs as a result of bacterial infection and/or injury to the mammary glands. At the onset the udder will be pinkish and hot and the teats may be pink or light blue. Wash and disinfect all equipment and the nest box, and for treatment to be effective (100,000 units of penicillin injected in the thigh) it must be carried out at this point. Death may result if treatment is not given promptly. Let her nurse her young if she can or will, but do not resort to a foster mother or the infection may be spread to her.

In the case of orphaned litters or when a doe won't nurse, you can foist some on another doe if they were born within three days of hers. To prevent her from rejecting them, put some Vaseline on her nose. In cases where you have no other doe to pass orphans or "extras" from a large litter on to, you *can* raise them yourself but, please, think twice. Raising any animal from birth is a taxing job, and rabbits are among the most demanding. They must be fed with an eyedropper or doll bottle at least four times a day (and at night). If you're not sure you'll stick with it, drown them and dispose of them. (You can feed them to your pig if you're up to it.) This is preferable to stringing them along for a few days slowly starving, and then deciding the drudgery is no longer worth it.

If you do decide to raise them yourself, use cow's or goat's milk, or evaporated milk mixed with water to the consistency of whole milk. Heat the milk so it is warm on the back of your hand and feed it with an eyedropper *at least* four times a day. (Eight times a day is preferable for the first four days.) At two weeks they can be offered rolled oats and a few blades of grass. By 15 to 18 days they can be taught to drink from a saucer and eat regular rabbit feed. Moistening the feed slightly will make it more palatable.

It is possible, even after you witness mating, that the doe will not conceive. There are a variety of reasons: sterility in the doe because of age or a long period of nonproduction, sterility of the male, disease, molting, she is not in heat (i.e., the four days when no eggs are available), or because of false pregnancy. False pregnancy occurs when a doe is mounted by another doe prior to breeding. This ruptures the egg follicles but obviously no conception will take place. False pregnancy results and conception by a buck is impossible for 17 days. This is a good reason not to house does in the same cage.

Because of the above cases of nonconception, it is helpful to be able to tell if a doe is pregnant earlier than the 31st day, so you can adjust her feed and rebreed her as soon as possible. You can feel the young most easily in the doe on the 14th day after breeding. (After that they are too large, and indistinguishable from the internal organs by all but the most expert.) It is best to test her in the hutch so as not to unduly upset her. Restrain her with one hand, holding the ears and the scruff of the neck as in force mating, and with the other hand gently feel the abdominal area immediately in front of the hind legs. At this stage the young are about the size of marbles and with some experience should be readily detected. To practice, use a doe you know is not pregnant and compare her with one you are sure is. In time the testing will be routine. A less reliable method is to reintroduce her to the buck in two weeks after the first breeding. If she growls or whines and otherwise rejects the buck, *chances are* she is pregnant. But this is not a sure thing—remember, even if a doe is not pregnant she may act the same way if she isn't in heat.

If a doe appears not to be pregnant she should be bred again. Even if you're sure she isn't pregnant and she accepts the buck again, (pregnant does have been known to be mounted even while pregnant) put the nest box in a few days before the original kindling date just to be safe. It won't hurt, and people have had litters from a "dry doe."

Rebreeding

In the wild, rabbits come into heat only in the fall and spring. With domestic rabbits, as mentioned earlier, there is a 4-day nonfertile period in each 16-day cycle, but otherwise conception can occur any time during the year. The most normal time to rebreed is after a kindling and eight-week nursing period. You shouldn't wait too long between weaning and mating as sterility may result. The rebreeding schedule mentioned above would enable a doe to have four litters a year, climate or warmth of your hutches permitting (31 days gestation + 56 days nursing = 87 days; 87 × 4 = 348). If an insulated or heated nest box still doesn't allow you to have successful winter kindling in your part of the country, you will be limited to a maximum of three litters a year: the first to kindle in early April (or earlier if your weather warms up before then) and the third to kindle in late September or early October.

Rebreeding can be carried out as early as the fourth week after kindling (simply remove the doe to the buck's cage as in normal mating), allowing a possible maximum of six litters per year, but this depends on your doe. Don't try to force her or she won't last very long. If she is healthy and chunky, you can rebreed her early; if she is drained and depleted, wait at least until she's weaned the litter or until she is back in condition. With experience and the daily handling of your rabbits, you'll be able to determine what is best for your herd and your does. In the case of a lost litter, the doe's feed should be cut to regular maintenance ration and she should be supplied with roughage to help her dry up. She can usually be remated three days later. When a doe has only a couple surviving young, they can be given to another doe and she can then be dried up and rebred in a day or so.

Experiments in breeding can be fun and rewarding if you begin to markedly improve your herd through selective breeding. An acquaintance of ours reported success with breeding purebred medium-weight breeds (New Zealands, Californians) to a Giant breed such as Flemish or Checker to produce larger fryers maturing in the same time as pure medium-weight rabbits. Experiment.

Bucks

Your buck is half your herd and you should choose him with that in mind. Look for all the qualities you do in your young stock. In addition, look for one with a shiny coat and bright eyes. Weight is not as critical as it is with does, but one that is overly fat may be lazy and not service your does. When that happens put him on a diet and his interest should return. The scrotum should be full and large and contain two fully descended testicles. If a buck has small and withered testicles (most common in older rabbits), he may be sterile or sire small litters. Bucks of the medium-weight breeds mature in six to eight months, but it is best to wait until eight months of age before using him. The larger breeds don't mature until at least ten months. The first few times, the buck may benefit from forced mating, especially when older does are used. As a rule of thumb, a buck can service up to ten does. Ideally he should not be used more often than every four to seven days.

HEALTH

Rabbits are among the most disease-resistant of livestock, *provided* they are well fed and live in clean surroundings. Prevention is once again the byword. Frequent cleaning and disinfecting of feed and water dishes and removal of spoiled hay, cleaning the hutch, and frequent removal of manure from underneath the pen are all good practices. A solution of lye is a good disinfectant. Blowtorches, commonly recommended for disinfecting, are not totally satisfactory because they must have direct contact to kill germs and also prolonged use may weaken the galvanizing bond on wire and invite rust.

Cleaning nest boxes after removing from the hutch and drying them in the sun is recommended. Sick rabbits should be quarantined for two weeks after symptoms disappear. New rabbits entering the flock should be similarly isolated to prevent them from bringing in any diseases. The table in the *Appendix* will be useful in the diagnosing and treatment of the most common rabbit afflictions.

BUTCHERING

While some may reason that you should wait until a rabbit matures to ten pounds before butchering, this is an unsound practice. The most efficient (hence most economical) growth occurs during the first eight to ten weeks of life, and thereafter feed conversion is not nearly so efficient. While you might get more meat, it is much more expensive.

Once you butcher a rabbit you may never again consider eating your own chickens. The meat is delicious, better for you and there is very little waste (20 percent). No more hours of plucking, feathers up your nose or in your shirt. Once you get the hang of it, you can butcher ten rabbits in the time it would take to do a couple of chickens.

Before killing, a rabbit must first be stunned. There are two ways of accomplishing this. One is to stun with a blow on the head and the other is to dislocate the neck. Dislocating the neck is faster, but I find it rather distasteful. To stun, grasp the rabbit in one hand by the loin and with the head pointing downward, deliver a sharp blow between the ears at the base of the skull with a hammer handle or other stick. In dislocation, hold the rabbit's legs with one hand and place the thumb of the other hand on the neck just behind the ears and extend the four fingers under the chin. Stretch the rabbit by pushing down on the head, press down with the thumb and rapidly raise its head to dislocate the neck. This method may need a bit of practice to get it right, and I'm unwilling to go through the ghastly near misses to perfect it. The squemish can shoot downward at the base of the skull with a .22.

Whatever your method, the head must then be severed and the carcass hung for a thorough bleed. Hang it by one leg after

Positioning of rabbit prior to dislocating neck

making a slit in one of the tendons. Cut off the free rear foot and both front feet. Remove the tail and carefully slit the skin (being careful not to cut the flesh) from the hock of the remaining foot down the inside of the leg to the base of the tail; continue up to the stub of the other leg. Now, carefully pull the skin down off the body, inside out like a glove. A knife may be used to separate the skin from the fat, which should remain on the carcass. (This process is almost identical to skinning a lamb—except the lamb is hung by both legs—for those of you familiar with that process or desirous of learning it.) The belly is then slit from the middle of the breastbone to the tail. Remove the entrails and bury or use them for stock feed. Save the heart, kidneys and liver, being sure to cut the gallbladder from the latter. Cut off the remaining foot and rinse with cold water to remove blood and stray hairs. As with most meat, cutting up is facilitated if the meat is well chilled, although it is not absolutely necessary.

A rabbit carcass, cut up, should have seven pieces. Remove the front legs and then the back legs at the ball joints. The back is then cut into three pieces, much like a chicken.

The hide is worth nothing commercially, but as mentioned earlier can be put to use around the homestead. Tanning can be carried out by following the directions outlined in the *Appendix.*

While eight-week-old rabbits are most often butchered, those older rabbits you cull can be made use of. Bucks are generally tougher than does but they can be stewed or ground and added to pork for sausage.

CHAPTER THREE

Sheep

While sheep may not rival many animals in intelligence, they can teach them a thing or two about feed efficiency. Like goats and cows, sheep are ruminants or multistomached animals which through bacterial action in their stomachs can convert cheap roughages and other inexpensive low-protein feeds into high-quality foods that are rich in protein, amino acids and essential vitamins.

Sheep are hardy, sure-footed animals that can graze even the hilliest and rockiest pastures unsuitable for other livestock. Because they eat a wide variety of roughages they are excellent weed killers and this, along with their rich manure, makes them an ideal animal to improve poor pasture. (They also make superior lawnmowers—much better, I think, than those outdated models that need gasoline.)

Lambs purchased in the spring will fatten quite nicely to butchering weight by the fall on grass alone. They need relatively little grazing space, and by using a tethering system and being careful, a few sheep can do well without any fencing. Sheep will also supply you with another valuable product besides meat: wool. In this Acrilan-orlon-nylon polyester age of ours, many of us have forgotten (or never knew!) the feel of real wool. A sheep will supply from 3 to 18 pounds of wool per year that can be used for spinning, knitting or sale to help defray the costs of your sheep and other livestock. Purchase a few ewes and a good ram and you will be supplied with lambs year after year, their cost being only a fraction of the cost of your ewe. By starting your own mini-flock, delicious lamb and useful wool will be yours for a lot less than the consumer pays.

BREEDS

There are many breeds of sheep, classified in six groups as listed in the table below.

TABLE 1: Classes and Common Breeds of Sheep

Fine-Wool Type	*Medium-Wool Type*	*Long-Wool Type*	*Crossbred Wool Type*	*Carpet-Wool Type*	*Fur Type*
American Merino	Cheviot	Cotswold	Columbia	Black-Faced Highland	Karakul
Debouillet	Dorset	Leicester	Corriedale		
Delaine Merino	Hampshire	Lincoln	Panama		
Rambouillet	Montadale	Romney	Romeldale		
	Oxford		Targhee		
	Shropshire				
	Southdown				
	Suffolk				
	Tunis				

We will be most concerned here with the medium-wool types. They are primarily raised for their meat, being rather blocky in conformation and relatively fast-maturing animals. They are not usually known for their wool production, but recently more emphasis has been placed on this in their breeding, and they are good dual-purpose sheep. If you are most interested in wool, you should look into fine-wool breeds which are also becoming better meat producers due to selective breeding. Long-wool types are generally too slow maturing to be important here, but crossbreeding them with the fine-wool breeds have made them a fair dual-purpose animal, and for that reason they are often classed with the medium-wool types.

If you are simply planning to purchase grass lambs (spring-born lambs that will be raised on pasture and butchered in the fall), breed will be of little importance to you. The key factor will be availability in your locale and personal appeal; although it is important to note that some breeds are faster-maturing than others and will lend themselves more readily to a spring to fall rearing program. If you plan to do your own breeding, a deeper knowledge of breeds will be necessary. In any event, some familiarity with the common breeds is called for.

Medium-Wool or Mutton Breeds of Sheep

Cheviot: White faces and legs covered with short white hair. Black noses, lips and feet with short, erect ears. They are quite hardy, prolific (125 percent lamb crop*) and they make good mothers. They are one of the smaller breeds and mature much more slowly so are not ideal for a summer raising routine. Light fleece (six to eight pounds). Mature rams average 175 pounds; ewes, 125.

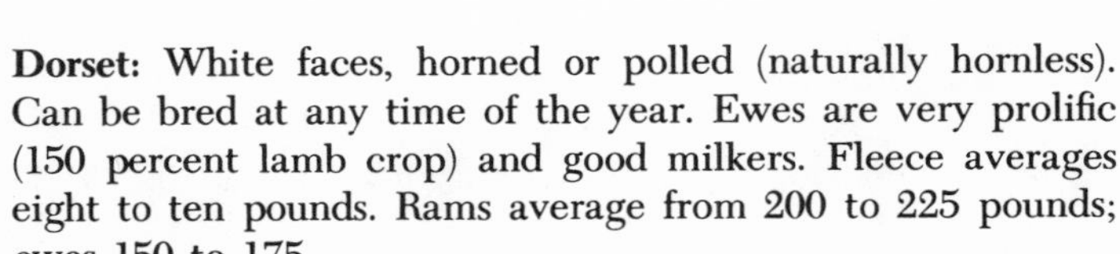

Dorset: White faces, horned or polled (naturally hornless). Can be bred at any time of the year. Ewes are very prolific (150 percent lamb crop) and good milkers. Fleece averages eight to ten pounds. Rams average from 200 to 225 pounds; ewes 150 to 175.

Hampshire: Black faces, ears and lower legs. The ewes are prolific and good mothers. This is a large fast-maturing sheep, and for that reason is excellent for summer raising, especially in cooler regions where the pasturing season is short. Fleece

* Lamb crops are commonly measured in percent as an average of lambs weaned. 100 percent would indicate one lamb per year weaned; 200 percent would mean twins weaned each year. 150 percent is the mark most breeders shoot for.

averages seven to eight pounds. Rams average 250 pounds, ewes 180.

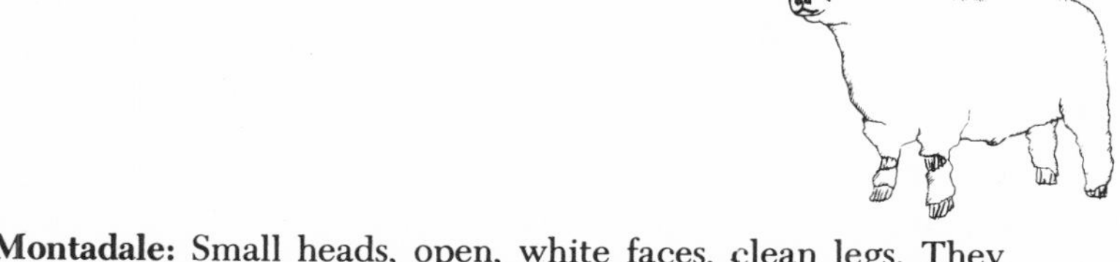

Montadale: Small heads, open, white faces, clean legs. They are prolific and good mothers. They are a very good dual-purpose breed with heavy (10 to 12 pounds) fleeces.
Oxford: A large sheep with gray to brown faces, ears and legs. A fast-maturing breed whose ewes are prolific and good milkers. Rams average 300 pounds, ewes 175 to 250. Heavy (10 to 12 pounds) fleece.
Shropshire: Deep brown or black feet, noses, ears and legs. Wool covering faces often results in wool blindness, which is a drawback, but recent breeding is aimed at eliminating this. A good combination wool/meat breed that produces a large carcass and a fleece that averages nine pounds. Rams weigh 175 to 200 pounds, ewes 135 to 150.

Southdown: The most desirable meat breed. Blocky and short-legged with gray to brown faces. They are small but mature very early. A light fleece (five to seven pounds) and rams weigh 175 pounds, ewes 125.
Suffolk: Open, black faces, black legs and ears. The ewes are good milkers and very prolific (150 percent lamb crop) and have a fleece weighing between eight and ten pounds. They are a large sheep, rams averaging 250 pounds and ewes 180.

Tunis: This is a hornless, open-faced sheep with a tan or red face. The breed is fairly rare in this country. Good milkers and lambs are early-maturing. Rams weigh 150 pounds, ewes 110 to 125.

It is not really desirable to purchase purebred sheep, except perhaps in the case of a ram (which will be discussed later). They cost more and they will not grow faster than good-grade sheep and will actually grow more slowly than crossbred animals. As with other animals, they lack the hybrid vigor associated with crossbreeds and as a result are not as hardy and have a lower lamb survival rate.

PURCHASE

In choosing a breed to start your flock with you should be influenced by the following factors: (1) personal preference; (2) availability; (3) products desired (wool or meat); and to a lesser degree (4) weather and other environmental conditions. You will do best to choose a breed that is pleasing to you over other factors since it will not be easy to sustain interest in a breed that is unattractive to you. If you want both wool and meat, one of the medium-wool breeds should suit you. Interest in sheep for wool only and not slaughter would lead one to choose from a breed where wool production is higher. If you have sparse or rocky and hilly pasture, you should consider one of the hardier breeds of sheep listed. In areas prone to severe and snowy winters and where undergrowth is thick, closed-faced breeds will be more prone to wool blindness.

It is best, if possible, to avoid livestock auctions and private individuals in buying your sheep unless you are truly expert at selection. Your best bet is a reliable sheep farmer. He will be better able to understand what you are looking for and should have a greater number of animals to choose from. He can also be very instructive.

You will be looking for all or some of the following: grass lambs to raise over the summer for butchering in the fall; ewes for breeding; and/or a ram or rams for breeding.

Grass Lambs

Buy your grass lambs as soon as you can after the grass becomes edible in the spring. This will give you a longer pasturing season and more lamb for your money. You will be offered ram lambs or wethers (castrated males); ewe lambs generally are not sold for meat purposes but are reserved for future breeding and are higher in price. While over a few years a wether will weigh slightly more than a comparable ram, for the purposes of raising over the summer differences in carcass and taste are nil. If you have breeding stock, however, get wethers or do the castrating yourself, or you will have frequent headaches trying to keep your meat lambs from breeding with your ewes.

Make sure that your lamb's tail is docked (painlessly cut off), since this insures cleanliness and helps prevent infection or infestation with maggots.

Look for: Large well-formed lambs that are straight-backed and stand squarely; deep chest, good width between front legs, well muscled and firm; alertness, and for one that walks with ease. Part the fleece and look for bright pink skin. The mucous membranes around the nose and mouth should be bright red.

Avoid: Generally the opposite of above: small, listless or generally unhealthy-looking lambs; or those with diarrhea, colds or runny noses. Avoid poorly proportioned lambs and those with dark, bluish skin or those with crooked legs or which exhibit lameness. A sheep whose front legs appear to emerge from one cavity may indicate inbreeding.

Note: In adding grass lambs to your current flock (i.e., older breeding ewes and possibly a ram), be advised that rather savage bullying of the lambs may result. We have discovered that adding grass lambs in pairs, at least, gives the newcomers some respite and helps alleviate the problem.

Breeding Stock

Ewes: In purchasing your breeding stock considerably more care is called for. While a grass lamb is raised for perhaps six months and slaughtered, a ewe used for breeding will be around for many years and affect your entire flock. You can buy a ewe lamb in the spring to raise for breeding, or purchase older ones in the late summer or early fall. Especially for the beginner, it is advisable to buy a ewe two or three years of age that has already lambed. Such a ewe will be mature and have had lambing experience, which will mean fewer demands on the owner. If possible, try to buy one that is already bred so you can start off the first year without having to purchase a ram and worry about a successful insemination.

Again, avoid livestock auctions and private owners who may try to foist their sickly ewes or poor lambers on you. A sickly ewe can barely support herself much less carry and milk a lamb. We've had good bargains buying culls from a large sheepman. These are not culls in the usual negative sense, but normal, healthy sheep from a very large flock that do not compete well for food in large groups. In our small flock they thrive. What is a loss for an owner of a large flock is your gain at a bargain price.

In general, look for the qualities you want in a good grass lamb. Look also for relatively large ewes because large animals tend to have larger, faster-gaining lambs. Twinning is inherited, therefore a ewe that is a twin, or has produced twins, is more likely to do so again. Avoid shallow-bodied animals or those that are long-legged or with narrow heads. Again, these points are relative and experience will help you to judge your stock. The fleece should be uniform, compact, bright and clean; avoid those with uneven coloring. The sheep should walk well and the feet should be well formed and uncracked. Foot problems will affect how well a ewe can carry her lamb. Avoid those with foot rot and those that are limping.

In large flocks, sheep of the medium-wool breeds don't begin to decline physically until about six years of age. Older sheep cannot feed well when grouped with younger, stronger ones. As explained earlier, older sheep that are culls from larger flocks can be a good addition to your small flock if the price is right. With the proper care given by a conscientious stockowner in a small flock, they can usually lamb for many years after they are "washed up" in the larger flocks.

Avoid young ewes (two to three years) that haven't lambed yet, for chances are that if a ewe hasn't lambed successfully by that time, she never will. These ewes are most commonly sold at livestock auctions and by disreputable people. With practice you will be able to spot such animals. If the ewe's teats are very small (relative, of course, to her age) it is probable that she hasn't had a lamb—or at least one that's lived so that she could nurse it. If there is any doubt in your mind, don't buy.

With all this discussion of age in buying, it would be helpful to know how to determine it. If the sheep is a purebred with papers, the task is quite simple. If there are no papers, the development of the teeth is a fairly accurate, albeit not infallible, indicator of age. A lamb has narrow teeth corres-

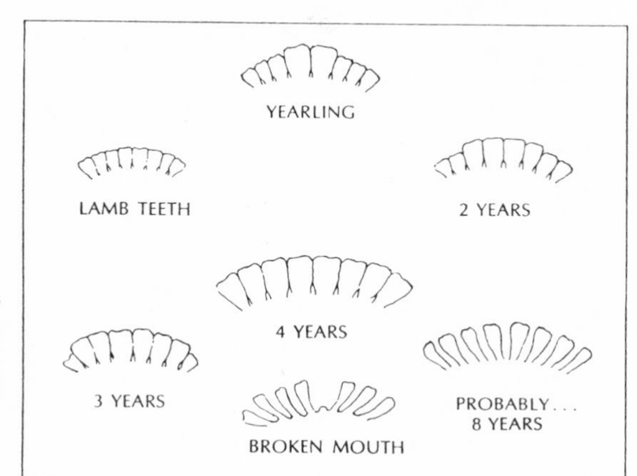

Age determination by stage of teeth development (see text). (From The Production and Marketing of Sheep in New England, a New England Cooperative Extension Publication)

ponding to baby teeth in humans. At the age of 12 to 18 months the centermost teeth are dropped and replaced by two broad permanent teeth. Each year thereafter an additional pair of permanent teeth appears until, at age four, the animal has what is known as a "full mouth." After this the determination of age by teeth is less accurate. Often the full set of teeth is intact until the age of eight or nine, but more often a few teeth are lost around age five or six resulting in a "broken mouth." Older sheep with no teeth are known as "gummers" and are not (no surprise) in high demand. With experience, the approximate age of older sheep can be determined by the degree of wear or spread of the teeth and the number of teeth lost. In a sheep with a full or broken mouth, fitness and age can further be determined by the condition and placement of the rear teeth. By feeling to the rear of the mouth (caution: sheep have strong jaws; have another person hold the mouth open while you do it) one can check the condition and shape of the rear teeth. Those with teeth that are still intact and meet well are apt to be of producing age, while those with missing teeth and abscesses should be avoided.

Perhaps the most important single consideration in the purchase of a ewe is her udder and teats. A ewe, no matter how many twins she drops, won't be worth a damn unless she can milk them adequately. The udder should be soft, pliable and free from lumps. Abscesses or ruptures of the udder and blind or missing teats, or those that are abnormally large or thick, should preclude your buying such an animal.

Rams: In choosing a ram spend even more time, often beginning to look many months in advance so that you'll find a suitable one and not get stuck with a poor substitute. For while a ewe passes her traits on to one or two lambs per year, your ram will affect every lamb in your flock. In ewes, you should get good grades or crossbreeds, but with a ram it is advantageous to spend a few extra dollars and get a purebred. In this way you'll be constantly upgrading your flock and imparting hybrid vigor. At worst, use a grade or crossbred ram that embodies as many desirable qualities as you can find (rate of gain, size, twinning, etc.), but remember that you won't be upgrading the quality of your sheep. If you breed with a ram that is on the whole better than your flock, you will in the long run get a better flock; by breeding to a ram that is equal to or poorer in quality you will downgrade your flock or at best simply not improve it.

In addition to the desirable qualities mentioned earlier in this section, select a ram that is large and thick with plenty of bone. Be aware of inherited traits listed in the chart below (which is also applicable to ewes). Choose a ram that is "full of it," that is, active and *ram*bunctious. It should be between one and six years of age, with strong legs and wide at the ears. Also, last but most certainly not least, make sure your ram has two fully developed testicles. It helps.

*Heritability Characteristics in Sheep**

Meat Characteristics	*%Heritability*	*Comments*
Twinning	15	Twins increase pounds of lamb sold.
Birth weight	33	Larger lambs at birth usually make faster gains.
Weaning weight	33	Weaning weight is tied to cost of production per head.
Yearling weight	43	This is the ability to gain on own without dam.
Post-weaning daily gain	71	Gain without mother's milk is efficiency of animal to produce on own.
Efficiency of gains	15	Pounds of feed per pound of gain important in economy.

Body type	12	Not highly heritable, but affects purebred value and market value.
Condition score	12	Important in lambs sold at weaning.
Wool Characteristics		
Face covering	43	Ewes with open faces produce 11.1 more pounds of lamb per ewe than closed faced ewes.
Neck folds	35	A defect in certain breeds.
Grease fleece weight	47	On average, 33 percent of income from sheep is from shorn wool.
Staple length	45	Major factor in determining fleece weight.
Fiber diameter	57	Important in determining wool grade.

HOUSING

Sheep are perhaps the most defenseless of farm animals, and fencing and housing are as much to keep predators out as to keep the sheep in or to protect them from the weather. If you plan to raise only grass lambs and not winter or breed them, and your area is relatively free from stray dogs and other predators, you're in luck. You can get by with tethering your sheep and providing them with shade in warmer weather.

Tethering

In tethering you attach a chain to a collar on the lamb and to a post with a swivel on it. With a swivel the lamb can move about with a minimal chance of wrapping the chain around the post. Lately I have seen large screw anchors intended for dogs that screw in flush with the earth. This will serve the same function as the stake but eliminates the biggest danger in tethering: the lamb wrapping itself around the post and choking.

° Prepared by Byron E. Colby, Veterinary and Animal Sciences Department, University of Massachusetts; quoted in part from *Genetics of Livestock Management* by John F. Lasley and *The Stockman's Handbook* by M. E. Ensminger; and made available by the Vermont Extension Service in cooperation with the Vermont Sheepherders Association, 1975.

We live on a road with little traffic and no stray dogs, and the first year we had two sheep we were able to leave them loose most of the time. The handful of grain kept them hanging around every morning but they never wandered far and put themselves to bed in the barn every night. This "honor system" worked quite well until a friend's child chased the lambs one day, and they came to the realization that they could in fact go anywhere they wanted. From then on tethering was a must, but with most sheep you can tether one and the flocking instinct will keep the other in sight. With a larger number of sheep the mathematics become more complex. One summer with five sheep we could tether three and keep the other two around; but if we tethered only two the three would wander off. Apparently sheep can add and subtract.

As we discuss sheep wandering off it is worthwhile to recommend selecting one lamb (perhaps the most rambunctious and thus the most likely to escape) as the "bell lamb." If they wander you can hear the bell at some distance. Try finding a lamb without a bell in the woods and you'll see what I mean.

Words of caution about tethering: it is not a system in which you can just leave your lambs unattended on the "back forty" until butchering time in the fall. It works best when they are close to the house and you can see them easily—one of the pleasures of sheep. They must have shaded spots to retreat to during hot weather (and be sure there are no obstacles such as trees, heavy weeds, briers, etc., around which they may become tangled). If you are in the habit of tying them out during the day and locking them up at night as a precaution against predators, be aware of the fact that during hot weather they often eat nothing during the day, choosing to graze at night when it is cooler. If you enclose them at night they will gain slowly or even lose weight. With tethering you must be on guard against stray dogs making quick work of a defenseless lamb, rendered even more defenseless by tethering. If dogs are a problem, you'll probably have to go on to more elaborate (and expensive) methods of containment.

Check the lamb's neck periodically for any sores or chafing from the collar. If not treated they may become infected and infested with maggots. Most of all, check your lambs daily and unwind their chains so they do not become wrapped around the stake. A word to the wise: tethering is a very good and cheap system for a few sheep *providing you are vigilant.*

Woven Wire Fencing

Sheep fencing is the most disagreeable thing about sheep. It is expensive. I wouldn't even consider fencing large areas unless I were going to do breeding and winter sheep. We've tethered up to five sheep at a time and that was our limit. It was more than we could safely handle—in fact one hung itself on a tether that summer. At the least you might buy a small roll of woven wire fencing to confine (and protect) your sheep at night or when you go away. Another option is to move that roll of fencing around as the sheep graze down one area. This is fairly simple, and will provide confinement and protection relatively cheaply.

For permanent fencing using woven wire, we have what I term "conventional" and "deluxe" systems. The conventional fence, and one that will do nicely in most instances, is 39- to 48-inch woven wire attached tightly to stakes so dogs can neither get under it or over it. If you still have problems with dogs or your sheep get out, the deluxe system, patterned after the USDA design, has 32-inch-high six-inch-stay vertical woven wire fence with one strand of barbed wire running along the bottom and two to five strands at the top, giving the fence a total height of five feet. This will keep out any dog, most people, and perhaps a charging elephant, and will cost accordingly. If you really want to go into sheep farming, this is the fence for you, but most people will get by with the conventional setup. Young lambs can sometimes slip through the holes in this fencing and may wander off. To prevent this make a triangular yoke out of lath. It should measure 8 to 12 inches on a side. This will prevent them from slipping through the fencing and will not hamper nursing.

Electric Fencing

Most people give sheep very little credit for intelligence, perhaps deservedly, but they can be taught to stay behind an electric fence. This can mean big savings for those who want to fence a large piece of land. It is considerably cheaper than woven wire and at least as easy to install. However, it does require frequent maintenance and checking, and if you live in a snowy region, will require resetting every spring.

For most purposes a two-strand fence of 12- or 14-gauge galvanized wire set at 12 to 14 inches and 22 inches will be adequate. Young lambs may still get under such a fence and if their wandering becomes a problem, you'll have to go to a three-strand fence set at 8 to 10 inches for the first wire, 26 inches for the top and split the difference for the middle wire. An electric fence will most assuredly keep dogs out and sheep in. The key factor is that the wire must be *very tight* (more so than for any other animal) or the wire will give and not push through the fleece to the skin. The fence charger should be the finest available and be correctly installed. As with any electric fence, it is imperative to check it periodically to make sure nothing is grounding it, thereby losing the charge (jolt).

Sheep should be trained before putting them in this type of fence to make them wary of it, prevent them from trying to get through it right away, getting entangled and shocked and breaking it in their panic. Training is best accomplished in the spring after shearing and ideally after a heavy rain that soaks both the sheep and the ground. Make an enclosure about three to five feet by five to ten feet, fenced with electric wire, and set the charger on it. Place a pan of grain in it and let the sheep in and allow them to attempt to get the grain. Watch them closely and be quick to repair the fence when or if it is broken down or grounded. Young sheep will learn more quickly than older ones who have never seen (or felt) an electric fence. This, and occasional jolts from their permanent fencing, should render them fence-wary until the next spring.

Shelter

If you are only going to raise grass lambs, you will have to provide shade for them, but no building or enclosure unless predators are a problem. For wintering sheep, especially in colder climates, some sort of enclosure is required. It can be simply a three-sided lean-to open to the south or southeast, or if early lambing is desired it should be enclosed and relatively draft-free. For later lambing, or in warmer climates, lambing can be accomplished outdoors and a lean-to will suffice. Do not cramp your sheep or force them to live in a filthy, foul-smelling pen. Allow at least 16 square feet of floorspace per ewe and furnish good bedding (sawdust, wood chips, straw, etc.) and keep the pen clean. Clean pens not only make for healthier sheep, but also make for better-tasting meat. The reason much commercial lamb has that heavy, "lamby" taste is that they are often forced to live in cramped quarters and sleep in their own excrement, which taints the meat.

Door openings should be at least eight feet wide to prevent sheep from becoming crowded in doorways and injuring themselves and unborn lambs. It is helpful to have electricity in the pen for a heat lamp in case of early lambing when warmth is needed. Lights also make working in the pen, especially during lambing, much easier. It is advisable that there be an outside yard off the sheep barn so that in winter the sheep, especially the ewes, can get exercise. We have our barn within the fencing and feed our sheep outside in the barnyard all winter. In this way they get plenty of fresh air and exercise.

Lambing Pen

If your ewes lamb when the sheep are back on pasture, no special equipment is necessary. However, if lambing is to take place inside you should be able to furnish a lambing pen. This is a portable enclosure that you use to isolate the mother and her lamb(s) from the rest of the flock for the first few days. This permits them to get acquainted and reduces the chances of a mother losing or abandoning her lamb. The pen should be about five feet square and must be in a draft-free location. If your barn is larger, as ours is, you may build some more permanent lambing pens. We use ours in the fall to keep our ram (or rams) away from the ewes until we wish them to breed. These pens can also be used to isolate sick sheep from the rest of the flock, to house a veal calf or some weanling pigs—or to confine a neighborhood child who will not quit chasing your sheep.

EQUIPMENT

If you don't have a stream or other supply of running water, you'll have to supply your sheep with a water bucket or a trough. Sheep don't drink as much water as most other stock, and much of their water needs in the summer are usually supplied by lush grass and dew, but a pail of water kept from the sun should be available at all times. When sheep are tethered they are apt to wind the chain around the water bucket and overturn it. If you're raising grass lambs, all you'll need is a water bucket and a small dish or discarded pan for feeding grain (if you feed any). For feeding hay and grain in the winter it is best to have a combination hay and grain feeder like or similar to the one pictured in *Figure 2*.

The first winter we kept sheep we fed them hay on the ground, and I would guess conservatively that a third of the

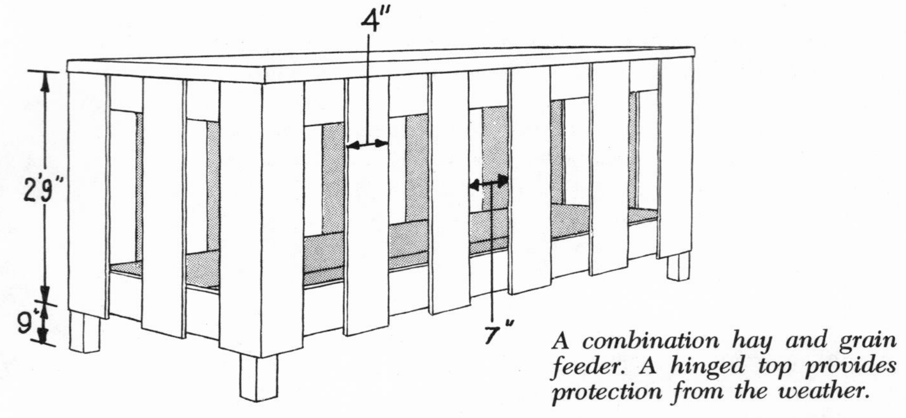

A combination hay and grain feeder. A hinged top provides protection from the weather.

hay was wasted: urinated on, trampled, etc. Although they do not quite equal goats in their refusal to eat off the ground, they will still waste a lot when fed that way. Since we began using a manger there has been almost no wasted hay. Whatever form of manger you use, remember the space between the slats that the head goes through should be seven inches so the head can fit in but not pull out too much hay at one time. The width of the manger is variable, but be sure to allow 15 inches for each mature sheep and 12 inches per lamb.

Some sort of box or dish should be provided, free from tipping and protected from the weather, to supply mineral salt to the sheep. It should be supplied in loose form, since salt blocks may chip teeth and subsequently hamper the ingestion of feed. (At times salt blocks are acceptable: see *Feed* section.)

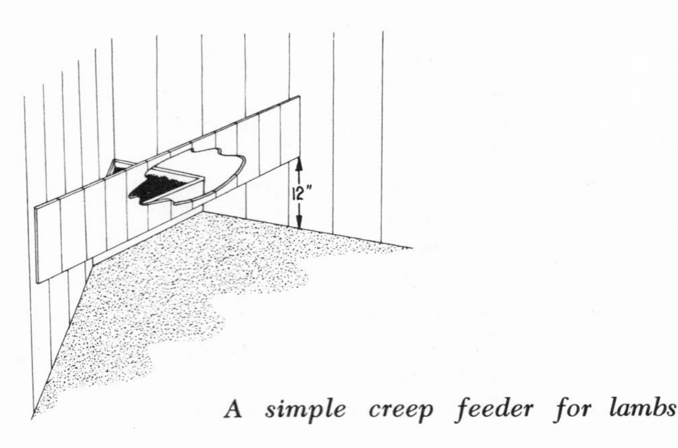

A simple creep feeder for lambs

If you wish to feed your lambs extra grain while young, which is often recommended, you will need a creep feeder. As with other animals we discussed, it is a feeder or a space that is only accessible to the young animals. The easiest method is to use an existing lambing pen or build one in the corner of your barn and construct the opening so that it is 12 inches from the ground and the lambs can get in at the grain and/or hay while the older sheep cannot.

FEED

Sheep are so desirable, especially in a food-starved world, because they can grow and produce meat, wool and lambs by deriving 90 to 100 percent of their nutritional needs from plentiful (and inexpensive!) pasture and roughages. As with the other multistomached ruminants (cows, goats, camels, etc.) they can take low-protein feeds and convert them into rich food. In the first of four stomachs, the rumen, the food is worked upon by millions of bacteria and microorganisms, and low-grade foodstuffs are transformed into essential protein amino acids. Also, many essential vitamins, chiefly the B-complex group, are produced and they need not be added to feed rations.

The rumen does not operate fully in a young animal, and so for the first few weeks of life a complete diet consisting of ewe's milk and/or a good creep feed must be supplied.

Grass Lambs

These rams raised for meat (or wool) only, and to be butchered in the fall, are the easiest of all animals for me to feed. I simply supply them with good pasture, plenty of water, salt and that's it. As a rule I do not grain any of my grass lambs. I do give them a handful now and then as "candy" to catch them or lead them somewhere, but none other than that. So far I haven't been convinced that it's worth it. Let us say, for example, that a lamb that costs you $30 in the spring will dress out at 30 pounds in the fall with no grain at all. That comes to $1 a pound for your meat. If you feed that lamb 100 pounds of grain concentrate over its lifetime (@ $8 per cwt.) and you realize an additional gain of 5 pounds, your meat then will cost you $38 for 35 pounds of meat or almost $1.09 a pound. Even if you add 10 pounds to the weight (which is unlikely) your price will drop to 95 cents a pound—a saving that is not all that remarkable. But do experiment for yourself. With the availability of supplemental feed in your region, you might be able to make it worthwhile. Consider, however, that if you breed your own lambs and they cost you $10 apiece, $8 (or 100 pounds of grain) will effectively almost double their cost.

If your lambs make slow gains or seem to be lagging, it could be that your pasture is inferior. In such a case, and in the fall when the pasture is dying, grain will be called for in the diet. I rarely feed more than one pound a day, but experimentation and suiting your own needs is the byword here.

Pasture: A good pasture can handle up to a dozen ewes and their lambs per acre. This is obviously a maximum and most pastures will furnish far less grazing. Watch your sheep carefully and note the condition and length of the grass. Sheep will graze very closely if allowed, and even kill pasture, so it is best to watch them to prevent overgrazing. If you don't have additional pasture you may have to give up a few sheep (or tether them on your front lawn) until you achieve a balance. The best system involves at least two pastures so that when one gets eaten low you can shift to the other and allow the first to regrow. This rotation of pastures is also important in the control of internal parasites. (This will be discussed in more detail later.) Sheep will improve a pasture in the long run, weeding, fertilizing and spiking in the fertilizer with their cloven hooves. A pasture that handles only four sheep per acre may handle six in a couple of years. It is also important to keep your pasture in shape with lime and additional fertilizing when needed. The best sheep pastures include alfalfa-brome grass, clovers, clovers and grasses, ladino clover mixed with other legumes and grasses and orchard grasses. Another good source of grazing is cut-over hayland after baling. In pasturing you must watch for certain plants that are poisonous to sheep, such as goldenrod, lupine and black cherry. Usually these plants will not be ingested unless pasture is very low and there is no other suitable forage available, but *veratrum* (false hellebore) is relished by sheep and can cause birth defects if eaten later in pregnancy. Burdocks can become hopelessly entangled in wool and make it worthless.

Water and salt must always be available to sheep. While loose salt is preferred over salt blocks because of less danger of injury to teeth, I have found that some of our sheep dislike the consistency of loose salt and prefer the salt lick. It is better to give it in that form rather than not have them get any at all. Two of the most important minerals for ewes are calcium and phosphorus, essential for the growing lamb and the production of milk. Often these minerals are missing from pasture and hays (especially if poor quality or washed-out hays are used) and must be supplied as a supplement. Always supply free choice in weatherproof feeders a mixture of either ½ mineral salt and ½ dicalcium phosphate* or ⅓ mineral salt and ⅔ steamed bone meal.

Important vitamin needs for sheep are A, B complex, D and E. If you are feeding good hay with at least ⅓ legume, your vitamin A needs will be met satisfactorily. Commercial grains usually contain vitamin supplements so check the bag if you feel deficiencies may be present. Other sources of vitamin A are cabbages, corn, squash and carrots which are all relished by sheep. As long as your sheep receive at least a few hours of sunlight a day their need for vitamin D should be met. Vitamin E deficiency will evidence itself by stiff-legged lambs and, if this is a problem, wheat germ oils should be fed and a veterinarian should inject your lambs with vitamin E. B-complex vitamins present no problems because they are synthesized in the rumen of sheep.

Hay: Hays of the type listed for pasturing will be suitable. If a high-quality legume hay or legume-grass mixture is used, it is possible to save substantially on feed bills by eliminating grain (see *Tables 3 and 4*). Second- or third-cutting hay, while usually a bit more expensive, is well worth it because sheep relish this finer hay and because fewer hard stems are present, less is wasted. Do not force sheep to eat the coarser hay by leaving it in the manger until they clean it up. It is less nutritious and may be removed for bedding or fed to pigs or other less-choosy stock. (We generally feed the remnants to our horses.) A mature sheep will consume 500 to 700 pounds of hay in a winter.

It is important that you not feed sheep a feed concentrate made up solely of corn. Also, don't put sheep out on very rich clover or alfalfa that is wet with rain or dew, or stomach bloat may result. As with all stock, feed on a regular schedule.

* This is not one of those villainous "chemical" additives to feed that are used by commercial growers. It is merely a naturally occurring mineral that is often absent from soils and must be supplied.

Ewes

Proper feeding of ewes will lessen stillborn births and make for healthier lambs that will gain faster. It will prolong the productive life of the ewe, increase her milk flow and lessen the chances of a ewe not accepting her lamb because of her weakened condition.

Flushing: As with other stock, flushing will improve your breeding record. Flushing two weeks before and up until breeding will tend to bring all the ewes in heat at about the same time and hence give you more uniform lambing times. It has been shown that flushing will increase your lamb crop by about 18 percent. Ideally, you will have a lean ewe gaining weight at the time of breeding. You do not want your ewes to be too fat or the result will be many unbred sheep. If your ewes are overweight switch them to sparser pasture six weeks prior to flushing so that by the time they are ready to be flushed they will be down to normal weight. Flushing can be accomplished by switching the ewes from fair to excellent pasture or if on good pasture or already on hay, adding ½ to 1 pound of grain to each ewe's feed per day. Upon successful breeding, feed as indicated in *Table 2* below.

It is just as important not to overfeed your ewes as it is to make sure they aren't underfed. Above all, ewes need plenty of exercise during pregnancy. Feeding them outside will make them exercise and also expose them to healthful sunshine.

TABLE 2: Suggested Daily Rations for Ewes in Good Health from Breeding until Six Weeks Prior to Lambing

#1	Grazing on good pastures may be continued, but care should be taken not to overgraze legume seedlings or they will be winter killed.	
#2	Barn feed legume hay	3-4 lbs.
#3	Grass hay	3 lbs.
	32-36% protein supplement	½ lb.
	or	
	Commercial mixed grain (12-16% protein)	1 lb.
#4	Mixed legume and grass hay	2¼-3¼ lbs.
	Protein supplement	¼ lb.
	or	
	Mixed grain	½ lb.
#5	Legume or grass-legume hay—good-quality	1½-2½ lbs.
	Corn silage	4-6 lbs.

#6	Corn or nonlegume silage	8-10 lbs.
	High-protein supplement	¼-½ lb.
#7	Good grass or legume silage	8-10 lbs.
	Hay	1 lb.
#8	Late cut timothy hay	3 lbs.
	Corn	½ lb.
	Oats	½ lb.
	Bran	¼ lb.
#9	Grass hay	3-4 lbs.
	Mixed grain (14-16%)	1 lb.

TABLE 3: Suggested Daily Rations for Ewes 4 to 6 Weeks before and up to Lambing

#1	Legume hay	3-4 lbs.
	Oats, bran, corn or barley (or mixture of all in equal parts by weight)	¾ lb.
#2	Grass-legume hay	1½ lbs.
	Corn silage	4-6 lbs.
	Mixed grain or bran and oats	¾ lb.
	High protein supplement	¼ lb.
*#3	Timothy hay	3 lbs.
	High-protein supplement	½ lb.
	Mixture of corn, oats and bran or commercially mixed grain	¾ lb.
#4	Mixed legume hay	3-4 lbs.
	Mixture of oats and bran or bran and corn or commercially mixed feed	1 lb.
*#5	Corn or nonlegume silage	6-8 lbs.
	Good nonlegume hay	1 lb.
	Protein supplement	¼-½ lb.
	Mixture of bran, corn, oats or oats and bran or commercial mixed feed	¾ lb.
#6	Legume hay	2 lbs.
	Timothy hay	2 lbs.
	Mixture of bran and oats, or bran and corn	1 lb.

Sheep need a grain that is between 14 and 16 percent protein. The next most imporant factor is its palatability to your flock. The generic names for these feeds may be sheep feed, goat feed, horse feed, etc. These are usually a balanced ration, containing vitamins, minerals and other necessary

* These rations may be too low in calcium. In this case or anytime nonleguminous roughages are fed, feed a mineral supplement as suggested on page 99. These tables are from *The Production and Marketing of Sheep in New England*, a New England Cooperative Extension publication.

foodstuffs. We have found 14 to 16 percent dairy ration to be both a relatively cheap feed and a very palatable one for our sheep. Experiment with different feeds, and find the one most favored by your sheep and your pocketbook. Good high-protein supplements include soybean, cottonseed and linseed oil meals.

You can make use of what feeds you have available to make up the most economical mix. Once you choose a program try to stick to it, and if you must change, do so very gradually. Be sure that silages are not spoiled or you are going to lose some sheep.

Up to four weeks before lambing you can save substantially on feed if you have your own legume hay. The less nutritious your hay, the more you will have to lay out for extra feeds. You can also save simply by not overfeeding your sheep. Up until a month before lambing the sheep should be well covered over the backbone and ribs, but you should still be able to feel bone in these parts. If you can't feel the bone, you're overfeeding. As *Table 2* indicates, you should increase feed a month or six weeks before lambing because 70 percent of the growth of the fetus takes place during this time. Again, follow recommendations because it is just as bad to overfeed during this time as any other. If overfed, the ewe will be too fat, the fetus will weigh too much and you will have lambing difficulties and possible loss of lambs confronting you.

When signs of lambing are imminent (see *Breeding* section), cut back on bulky feed and add ½ pound of bran to the ration as a laxative so there will be less competition for room between the end of the large intestine and the vagina at lambing time. This is especially important for first-year lambers and ewes that have a history of difficult lambing. After lambing, don't grain for 24 hours, since udder problems may occur and the lamb may scour. Then gradually build the grain back up to two to three pounds per day depending on how the ewe milks and the lamb reacts. In the case of twins, which are quite a strain on a ewe, heavy graining is critical. Other feed such as carrots, beets, cabbage or cull potatoes fed chopped and not to exceed three pounds a day are relished and are an excellent aid in the production of milk. When lambing occurs soon before or after sheep are put out to good pasture, as we do ours, we give mixed grain up to three pounds a day in addition to the pasture and taper it down to nothing in six weeks or upon weaning, depending upon the condition of the lamb(s) and ewe.

Creep Feeding: Sometimes, especially during early lambing (when pasture won't be available for weeks or months), or in the case of very young or old, rundown ewes, it is beneficial to creep-feed lambs. Sometimes, depending upon your beliefs, it's used simply to give lambs a boost to get them off to fast gains. Using the same sort of creep-feeding setup (see *Housing* section for details), feed very fine (second or third cutting) hay and finely ground grain from ten days of age on. After six weeks the grain needn't be finely ground, or when good pasture is available, the creep feeding can be discontinued unless you plan to grain your lambs the entire summer.

Breeding Rams

Generally, a ram will get by quite well on three to four pounds of good hay per day and no more. If he seems really weak or thin, I would give him a little grain—but he probably has worms or some other health problem. For two weeks before and during the breeding season I like to give him a 1/4 to 1/2 pound of grain per day for added energy.

Replacement Ewes and First-Year Lambers

I like to feed our replacement ewes a pound of mixed grain per day all summer long into the breeding season to give them the extra feed needed for faster growth and the rigors of lambing. Follow any of the programs outlined in *Table 2* for them, but supply one pound of mixed grain (or the equivalent) extra for growth. Remember, your first-year lambers are growing themselves as well as feeding a young fetus and you must take that into consideration. Do be watchful to see if they become too fat or rundown and adjust their feed accordingly.

The problem of feeding different rations to your sheep in the winter (e.g., young ewes needing more grain than older ones; ewes needing grain but rams not) is solved by large operators by putting sheep into different winter quarters according to what they eat. The small sheep owner wouldn't want the expense of different pens nor should he have to go to the bother with a small flock. Neither would he want to feed a ram two pounds of grain a day when he doesn't need it, or a ewe more than she needs because the younger ewes need that much. It is both expensive and possibly injurious to your ewes to overfeed. We have handled the problem by herding all our sheep into the barn and closing the gate. We then herd out those who get one ration, and keep in those that get another. This is admittedly very hairy at first, as *all* the sheep want to go out. But do this for a week and your "dumb" sheep will learn: herd them in, then open the gate and the ones that usually go out will do so, while those that are fed inside will stay in.

Supplementing Commercial Feed

Because of the "protein factory" in the stomach of ruminants, sheep can be fed cheaply on conventional feed alone. While grass lambs can be raised to butchering weight on grass alone, breeding ewes will generally need other food supplements besides grass or good-quality hay. The numerous feeding programs outlined earlier should afford stockowners a greater opportunity to feed their flock on what they can grow themselves, or what they can purchase cheaply in their locale.

As mentioned earlier, cull potatoes, cabbage, turnips, and carrots are a good supplement to regular feed and are especially good for milking ewes. In feeding these items do not exceed three pounds per animal. Do not make a change in the amount of grain or other concentrate you feed, but allow your savings to come in decreased hay consumption.

Look for special feeds in your area such as leftover bread at supermarkets and bakeries. Wheat and rye bread are about 8.5 to 9 percent protein and dry breadcrumbs are almost 12 percent protein. Graham crackers and saltines are 9 percent protein. Many supermarkets throw away leftover greens (make sure they aren't rotten) that the sheep like. Again, these foods cannot make up the total diet of the sheep but they can afford you savings in hay. As a precaution against underfeeding, in the case of the vegetables listed above, do not change the grain ration but allow them to eat less hay. In using supplemental feeding, the overall condition of the sheep, indicated by pink skin and the covering of meat on the backbone and ribs should offer you guidance. Experiment.

Two other supplemental feeds that I have heard about second-hand but have not tried myself are seaweed and a ration consisting of 75 percent hemlock (or other nonresinous evergreens) browse and 25 percent of a high-protein supplement. It was reported in the December 5, 1975, *Maine Times* that sheep kept on York Island ate seaweed in the winter they foraged for themselves. I have not experimented with this, but it may be a starting point for those of you who have access to seaweed. My concerns would be suitability for growing or pregnant sheep and the effect of the feed on the meat. Perhaps it would be more suitable for mature animals raised for their wool only. The same reservations hold for the feeding of hemlock to sheep. Avoid feeding the resinous, turpentine-bearing plants such as pines, firs and spruces as turpentine is well known for its abortive qualities. (C. F. Lucker reports that a neighbor of his fed a half-dozen discarded Christmas trees to his goats and all four lost their kids.) The sheepman who told me of this program has never done it for extended periods of time nor has he experienced any nutritional problems or loss of lambs. The University of New Hampshire has done research on deer browse (deer, although not ruminants in the truest sense, are nonetheless digestively similar to sheep) consisting of eastern hemlock, red maple, striped maple and hobblebush. These feeds were chopped up by a silage chopper and fed free choice at two feedings a day. The results showed that these feeds alone were not sufficient to maintain the deer, but other factors (most notably the confinement of a wild animal and subsequent drop in feed intake) could have affected the results adversely. It is still plausible that the addition of a high-protein supplement at a rate of 25 percent by weight would make it an acceptable feeding program, but that has yet to be verified experimentally. Unless more data are available, or you experiment carefully and find a good routine, it is best to limit browsing to what is naturally available in your winter pasture. Good savings are possible depending upon your pasturing conditions. Again, leave the grain portion of the feed unchanged and cut back on hay if they begin to leave more between feedings because of their increased intake of browse. As with all supplemental feeding, a good mineral supplement should be available, free choice, at all times.

MANAGEMENT

Routines

The type of routines you establish with your sheep will, of course, depend upon your demands and the size of your flock. If you are simply raising grass lambs, you will obviously butcher them all off in the fall and be done with sheep until the following spring when you buy a few more woolly little lambs. We have a ram and a number of ewes and after lambing in the spring we keep a few meat lambs, any replacement ewes we might need, and sell the rest. This extra income more than pays for the feed for the ewes the past winter and any surplus goes to lower the overall price of our meat and wool. Auctions are a quick way to sell surplus sheep or lambs, but generally a higher price can be fetched through private sales. You can usually expect to get 30 to 50 percent more for ewe lambs because of their breeding potential, but check average prices in your area in a market bulletin before selling. Any runts you may have on hand around Christmas and Easter can often be sold at a good profit (quite often higher than larger full-grown sheep) because of the demands then by certain ethnic markets.

Another bonus in sheepraising is your sheeps' wool and hides. Shearing is carried out in late spring or early summer each year. Depending upon the breed you may get 18 pounds of wool per sheep, but for the medium-wool breeds the average is 5 to 12. Your grass lambs can be sheared before butchering in the fall and the hides tanned (see *Appendix*) or, if you prefer, you can tan the hides with the wool intact for use in making slippers, gloves, rugs, coats, etc. Unless you plan to be snowed in all winter or you have a large family and can delegate work to your children, you probably won't be able to spin all your wool. There are a number of establishments that do custom spinning (listed in *Appendix*) and dyeing wool. You send your wool and you get the same amount back (minus a certain percentage depending on how dirty or full of burrs the wool is) dyed to your specifications. All this for 50 to 75 cents a skein! The wool from one lamb will keep you in socks for years.

We generally send all our wool to be spun, rather than selling off the surplus we don't need, and sell it by the skein. We charge far less than the store price for 100 percent wool but still a good deal higher than our cost which helps pay off the flock and possibly generate a little profit.

Some handspinners in your area might well be interested in buying your whole fleeces. Also, look into wool pools in your region. These are central points where you bring your wool, tagged and bagged, on a given day, and receive credit for the number of pounds of wool you brought. When the wool is sold you receive payments minus the cost of running the pool.

Keep signed receipts from the purchasers of the wool and at the end of the year file these at your local U. S. Agricultural Stabilization and Conservation Service (ASCS) office so you can participate in the wool incentive payment program. This program is financed out of tariffs on imported wool (not our tax dollars) and pays the seller a premium depending on the wool he has sold and the national average price for wool in this program. The ASCS office near you can give you full details.

Sheep are also beneficial for pasture improvement. As long as you don't allow them to overgraze they will do nothing but improve your land. Besides producing rich fertilizer, they also eat about 90 percent of all weeds, so after a few years you should have a lusher, weed-free pasture. Liming and top dressing with poultry or other available or cheap manure should also be carried out periodically. Lambs have been known as good flower-bed weeders, plucking out the nasty weeds and leaving most flowers intact. They do not, however, do as admirably in the vegetable garden, having so many favorite crops that just so happen to be human favorites too.

Handling

Your problems in handling, or retrieving loose sheep, will be minimized if you tame them early in life or soon after you buy them. One of their greatest defenses is their timidity, and they will shy away from most anything unless given reason not to. If you feed them by hand with some grain, petting them and talking to them often as you do so, they will begin to lose their natural fear of man. While this can be handy if any get away or during shearing, it can also be a great help during lambing if your ewes feel secure and unafraid around you. If your lambs have escaped their pens, a little grain in a dish will usually lure them back. Do not try to reason with them, they will not listen. In catching especially shy sheep we have found making a snag out of baling twine to be very effective. Make a slip knot and attach 10 or 20 feet of twine to it and sprinkle some grain around the loop. Be patient. When the sheep steps into the loop give a yank and you'll have your stray. In guiding stubborn sheep to a desired point, grab with one hand under the jaw (gently) and push with the other hand from behind. Steer with the head. If you really want to get fancy, you can buy a shepherd's staff from one of the farm supply companies listed in the *Appendix*.

Unless you have only a few sheep and they are easily told apart, you will need to mark your flock for identification. This helps immeasurably around lambing time and in your record keeping. A commercial ear tagger can be purchased for easy marking, but with a relatively small flock you can do just as well with a concocted system such as a bell on one, a collar without a bell on another, no collar on another, etc.

Predators

Dogs are by far the most common predators of sheep. Depending upon your part of the country, coyotes too may be a threat. While your sheep are at the mercy of a predator while tethered, a good fence (as discussed earlier) should preclude animals from getting in and killing all or part of your flock. Keep an eye out for any strange dogs in your neighborhood, since they are candidates for sheep killing. Be especially watchful when friends come to visit with their dog. Dogs that have never seen sheep before are naturally quite curious. It usually begins quite innocently ("They're just playing," are famous last words), with the dog playfully running after the sheep: then he gets all worked up, nips, and nips again, draws blood . . . and you know the rest. Prevention is your only recourse.

If you lose any sheep to dogs, local dog license funds usually provide for compensation for the owner. On rare occasions, sheep may be lost to bears, or other wild animals, and many states provide similar reimbursement from fish and game funds, providing you can prove loss by producing the carcass (which is often quite a trick because that is presumably what the bear came for in the first place).

BREEDING

If you want to get into breeding, the advice is, as with all animals, to start slowly. It's a good idea to buy a bred ewe or two the first fall or winter you go into breeding. This frees you from the need of a ram and concern as to whether conception has taken place. You might be able to get a good ewe, somewhat advanced in age and not a good competitor in a large flock, but which will do well in your flock. Our first ewe was one we bought bred in January for $30. She was perhaps six years old and doing poorly (although she was by no means sickly) in the large flock; no sooner was she in ours than she quickly established herself as the matron of the flock and to this day dominates all the other sheep. She has a few more years of lambing in her and she has already supplied us with lambs and more than paid for herself.

Most of the medium-wool breeds we are concerned with come into heat fairly regularly, beginning in the late summer or early fall and continuing (if not bred) until January. The heat period is when the ewes will permit the rams to mate. Its duration is 3 to 73 hours but three-quarters of the sheep stay in heat from 21 to 39 hours. The period will occur every 13 to 19 days (with the average being every 16½ days) until the ewe is bred or until the breeding season is over. The Dorset and Tunis and their grades or crosses can breed any time of the year and caution should be exercised so that they are not bred at inconvenient times. (Inconvenient, that is, for *you*!)

With sheep we have one of the few larger animals whose feed economy makes it possible to keep a male for breeding purposes even in a very small flock. The ram should be allowed to run with the rest of the flock the entire year except before the ewes begin to come into heat in the late summer or early fall. It's a good idea to be extra careful and not risk early pregnancies by removing the ram in plenty of time. In Vermont, we pull our ram in late August, and have never had any "mistakes." One ram from yearling age up until five years of age can handle 30 to 45 ewes, although it's a good idea not to tax him too much, and 30 is a good maximum (for really taxing breeding, feed the ram even more than recommended earlier, up to one pound of grain per day). A good ram lamb should be able to service 12 ewes without any trouble. After six years rams begin to lose potency, and if you find yourself with a large number of unbred ewes it may be time for a change.

The gestation period for sheep is 144 to 152 days with an average of 147 (21 weeks); therefore, about five months or so before you want your lambs, you should set your ram in with the flock. Before doing this it is important to carry out the following procedures: eying (ewes and ram), tagging (ewes), and ringing (ram). Eying involves clipping the wool from around the face and eyes if it is present (mostly in closed-faced breeds). This improves the eyesight and prevents eye irritation from the wool and seeds and dirt carried in it. Tagging is the clipping of the wool around the dock of the ewes. This promotes cleanliness and most importantly permits the ram to couple with the ewes more easily. Some people shear a ram completely prior to the breeding season, but in a cold climate such as Vermont we do not do this because the ram will not be able to grow a full fleece back before the onslaught of winter. Ringing is an acceptable substitute and consists of clipping wool from the neck and from the belly around the penis. This makes it easier for the ram to make proper contact with the ewe during mating. Excessive heat may affect the potency of your ram adversely, and if breeding is to be carried out in very hot months, complete shearing may be necessary to insure full potency. Your sheep should not be wormed during the month before they are bred.

As I mentioned earlier, letting your ram run with the flock all year (with the noted exceptions) is good because the ewes will know him and therefore be more likely to accept him. If possible, when you separate the ram try not to alter his feeding or grazing routine very much, if at all, since possibly this could affect his performance. Two weeks before setting your ram in with your ewes you can flush them and also, if possible, secure your ram in sight of the ewes (but still out of contact). This is called "teasing" and can help stimulate ovulation in the ewes.

Unless you are a ram, it is very difficult to tell when a ewe is in heat. Often they will begin mounting each other and their vulvas may be red and swollen, but this swelling is not nearly as pronounced as in pigs. You can usually tell if they allow the ram to mount them. In most cases, a ewe will not be receptive to the ram unless she is in heat. If you are lucky you may see your ram breed with your ewes, enabling you to calculate approximate lambing time from that date—*assuming* it was a successful insemination. (A word of caution: it is illegal to watch sheep copulate in 12 states and Puerto Rico. It would be a good idea to check your state regulations beforehand!) Or you can just leave the ram in and simply assume all your ewes are bred and have the lambs come as a surprise the following spring. A better system, in which you can be reasonably sure of the date of lambing, is to equip the ram with a marker on his chest so he marks the back of the ewe when he mounts her. You can buy a harness with a crayon attached from farm supply stores or you can mix a marking paste to apply to the ram's chest once a week or so. A simple paste can be made by mixing mineral or linseed oil (do *not* use motor oil) and any available nontoxic dye such as venetian red or lampblack. Whichever method you use, you should change the color of the crayon or paste every 16 days, using in progression yellow, red, and then blue or black. They are used in this order because they cover the previous color. The procedure: set your ram in with the first color. Within 16 days (the average heat cycle) all the ewes will come into heat and should be mounted by the ram. They will then be marked by the first color. The day you notice the mark on the ewe is approximately 21 weeks from their lambing time—*tentatively*. After the 16 days change to your next color. If none of the ewes is marked with this color, it means they are bred because they will not be coming back into heat, and they will not accept the ram. Any ewes not marked again can be considered bred and should lamb 21 weeks from the day of the first marking. Any ewes marked with the second color can be figured to lamb (*tentatively* again) 21 weeks from that day. The color will be changed again after that 16-day period . . . and so on. In small flocks such as we are discussing, all ewes should be bred in their first or second heat after being with the ram. If none of the ewes is bred, your problem is most assuredly the ram. If a few ewes are not bred, they may be too old, in poor health or sterile. They should be considered for mutton or for sale.

While sheep in commercial flocks are bred for only five years, those in small flocks generally receive such good care that they can lamb successfully for as long as ten years. As a sheep gets older its teeth deteriorate, hampering food intake, and in such a case it may not breed successfully, or may not be able to nurse the lamb, or it may simply be so physically taxed that it will die during pregnancy. Keep an eye out for a ewe having trouble. Sell, butcher, or keep her as a pet, living out her twilight years without the drain of children (but still supplying wool, of course).

There is some disagreement as to how young you can begin breeding a ewe. Some say they shouldn't be bred until their second year; others recommend breeding in their first year, to lamb at the age of 14 months or so. Recent studies have shown that ewes bred as lambs will produce a greater lamb crop during their breeding life than those bred as yearlings. If you are going to breed ewe lambs, they should be at least nine months of age at the time they are bred and well grown. They must receive special feed consideration and given greater attention at their first lambing, because ewe lambs have a greater incidence of lambing problems and a disowning than yearling or older ewes. There is a greater chance that you will lose your first lamb crop from ewe lambs, but they will nonetheless still have a more productive breeding life than if they had not been bred.

A week or two before you believe lambs are due, check to insure that the ewe's udder and teats are free from wool. If not, clip that area and around the eyes (tagging again) if necessary. On the day of lambing, a ewe *may* exhibit one or more of the following signs: she will be off her feed, restless, isolated from the rest of the flock, dig or paw at the ground or bedding while circling a particular spot. She will appear sunken in front of the hips; the vulva may be red and enlarged; and, of course, the udder will be substantially larger and the teats coated with a waxlike film.

When you notice any sign of labor, watch but leave her alone and don't upset her by constantly peeking and feeling her stomach. Sheep may need help at times, but they are not nearly so problem-prone as they are reputed to be. First-lambers tend to have more problems, but if your flock has been well fed and exercised, you shouldn't have any major problems. A friend of ours who has the largest flock in Vermont had 300 lambings last spring, and only had to assist in one. Remember to watch your ram to see that he doesn't injure the new lambs.

Lambing: What to Have Handy

A clean pail
Vaseline, mineral oil
Soap
Soft, heavy string
A lambing loop (plastic-coated electrical wire)
Uterine capsules
Dry rags
A helper

Do not leave pails of water where lambing is to take place because it is possible that lambs will be dropped into a pail and drown. Offer water to the ewes periodically and then remove the buckets. The normal position at birth is the backbone of the lamb toward the back of the ewe, nose presented first with front feet presented alongside it. If afterbirth is passed soon after the birth, generally there will be no more lambs. If afterbirth doesn't come and the ewe continues to show signs of labor, paying little attention to the lamb, another will probably follow. In rare instances, three lambs (or even more) may be born, but you needn't bother yourself about such occurrences. If the lambs are born out on pasture, you needn't worry as long as the lamb(s) seems healthy and the mother cleans it off and allows it to nurse. If born inside, cut the umbilical cord off at four inches and apply an antiseptic to prevent tetanus. If conditions are crowded, place the ewe and her lamb in a lambing pen for a couple of days. You can do this before lambing if she is obviously ready to drop the lamb. Make sure she can see the rest of the flock and doesn't get upset. This allows the ewe and her lamb to get acquainted without danger of losing each other—most important for yearling and ewe lambs. In moving the new lamb to the pen, grab it gently under the chest and back it into the pen keeping it under the ewe's nose. For at least the first few days the ewe knows her lamb by smell only, and if she or it wanders away and it mixes with other sheep, she may disown it.

In chilly weather some burlap or blankets hung around the outside of the pen will help keep out drafts. A heat lamp can be hung in the corner of the pen so that the lamb can get close to it by choice. To prevent ewe or lamb from getting burned, hang it three feet from the floor. Do not use it for more than three days, because the lamb might get chilled when removed from the pen. If all goes well, the lamb and ewe can be let out of the pen in two days and allowed to run with the flock. Follow the feeding instructions as listed earlier and you should have no problems. You can wean the lamb as early as four weeks or let it wean itself naturally. We generally keep all our sheep together and in time the ewe will dry up and not allow the persistent lamb to nurse. As the ewe dries up, or as an aid to help her, slowly taper off her grain supplement until her feed routine is back to normal.

Castrating and Docking

Lambs, both male and female, should have their tails docked (cut off) within the first week or two. In addition, you may want to castrate the ram lambs you do not wish to save for breeding. Castration of grass lambs makes no substantial difference in the quality of the meat. However, their rate of growth may be slightly slower than the uncastrated. The advantage of castration is that you won't have to separate the rams from the ewes when they start coming into heat (usually well before butchering time). Segregation is not an easy task.

When lambs are 7 to 14 days old, their tails should be docked and they should be castrated. Castration should come first followed by docking. The three commonly used tools are a knife, an emasculator or an elastrator. Use clean instruments and perform operations in a clean place. A sharp knife is the cheapest but the most difficult to use. The lamb is seated on its rump; the hind legs are pulled up toward the front legs, and a hind leg and front leg are held in each hand. (Obviously, unless you have three hands, this is why you need a helper.) One-third of the lower end of the scrotum is cut off and the testicles are forced out by squeezing them from the base of the scrotum. You then pull on the testicles until the cords break. Spray the wound with an antiseptic and be sure not to tell the lamb what he'll be missing. While this is the cheapest method, it does create an open wound that is more likely to get infected or infested with maggots. It is also the most difficult for the novice to perform. An emasculator is an instrument that crushes the cords leading to the testicles and they dry up and wither away. This is a good method because it is relatively bloodless, but carelessly done can cause an incompletely castrated ram. Also, emasculators are quite expensive, running about $50 to $100. An elastrator is an instrument that spreads a tight rubber band over the scrotum when it is released. This cuts off the circulation, and the scrotum and testicles gradually dry up and fall off—usually within two weeks. It is quite easy to do; the instrument costs about $10 to $15, and there is no open wound.

Docking can be accomplished by using any of the three tools described above. (A fourth method, docking with a hot iron, is not included because it cannot be used for castration.) Handling of the lamb is much the same as for castration. For ewes, the tail should be cut to about two inches from the body; on rams, the same length or slightly longer so that you can tell them apart. This cutting of the tails makes mating easier for ewes, makes for greater cleanliness and lessens the chance of maggots. In using a knife, locate the point of cutting and push the skin toward the body so there will be extra skin to cover the stub. Use a *sharp,* clean knife and twist the tail a quarter turn before cutting. This causes the blood vessels to be cut diagonally and lessens the chance of severe bleeding. After cutting the tail, pinch and hold the end to stop bleeding. This should not be performed on lambs past ten days of age, and ideally not over a week old. I did this to a friend's three-week-old lamb and it nearly bled to death. I sauntered cockily into their kitchen, grabbed a knife and promptly severed the tail of the lamb that was being held on the table. A stream of blood shot across the room, splattering books, the floor and the walls. We used a rubber band tightly wrapped around the stub as a tourniquet, otherwise the lamb would surely have bled to death. This is the simplest way (if done properly and *early*) of docking a lamb's tail, but watch for maggots around the wound and apply a commercial spray, best purchased from your veterinarian. Ordinary antiseptics will not control them.

The elastrator is used for both docking and castration. The only criticism of it is that the tail does not fall off right away and the stub may be prone to infection. Docking and castration in cool weather will minimize the danger of maggots. The emasculator may also be used for docking. With younger lambs it will sever the tail, but with older ones the tail may have to be severed outside the jaws of the elastrator with a knife. It will crush the tissues and main artery and prevent bleeding. All the above procedures have their pluses and minuses—price of equipment may be a factor. If you feel uneasy about any of these operations, watch them done by an experienced person before you attempt them. Keep an eye on all wounds until they heal to detect any infections or parasite infestations.

Problems During Lambing

Sometimes three to four hours are required during lambing. Examine the ewe a half hour after the first water bag is passed and note her progress. If problems develop, the novice or squeamish may want to call a veterinarian or other qualified person to assist. You may watch and learn from them. I personally like to do things myself and feel the best way to learn is by doing it. Have all those helping wash their hands thoroughly. If you enter the ewe with your hand, coat it first with a lubricant such as Vaseline. Make sure your fingernails are short and smooth, and if possible bend your fingers at the knuckle before inserting them to avoid scratching. Lay the ewe on her right side and probe with your hand, trying to determine if the lamb is in the proper position. Ideally, its nose should be between its feet. (You can see now the benefit of having your ewes tame and used to your presence.)

A few hints: to distinguish between the front and hind leg, feel above the knee. The hind leg will have a prominent tendon, the front leg, muscle. If the hooves of the lamb are pointed up (toward the backbone of the ewe), they are probably front legs coming in the normal position. If they are pointed down, they are probably the hind legs and the lamb is in a breeched position. The most common difficulty is when *one or both* front feet are folded underneath the lamb. In the case of one foot folded back, tie a soft string around the presented leg (so it will not be "lost" in this procedure) with plenty of string protruding. Then push the lamb back slowly and gently. By sliding the hand down the neck you can locate the leg that is turned back and straighten it out. The lamb is then in the normal delivery position and should be passed without further problems. If the ewe is too fatigued and cannot at this time pass it unaided, make a loop out of insulated wire and place it around the lamb's neck. String tied around the feet can be pulled along with the loop and the lamb should come. You can pull quite hard, surprisingly, without hurting the lamb. However, always pull *with* the birth contractions. In the case of *both front legs doubled back* try the same procedure as outlined above. If that does not work, place a loop of string around the lamb's head with the knot in its mouth. Elevate the ewe's hindquarters and gently force the lamb back until you can run your hand down the neck and flip the legs through the pelvic arch. If the head has not slipped back through the arch, be sure to start it through before you start the legs through. In all the above operations lubricant applied to the vaginal wall will make it easier to maneuver the lamb into position.

Often in the case of hard labor, the *head of the lamb swells* and makes lambing even more difficult. Sometimes you will be presented with an abnormally large lamb. In both cases, if it is in the normal position, work the legs forward a bit and pull them gently while using the other hand to pull the head from side to side with the fingers positioned behind the ears. A helper can stretch the top of the vulva to ease the operation.

In a case where *both legs are presented and the head is back* first, put strings on each of the feet (allow plenty of extra string) and push them back through the pelvic arch. Then try to work the lamb's head through the arch. If you work with the ewe's contractions, you should be able to get the head and neck righted. If the head keeps twisting to the side, use a lambing loop around the head and try to pull it through the pelvic arch. Then pull the leg strings, and the legs will flip through and the lamb will be in the normal position. If none of these procedures work, you will have to turn the lamb and present it in the breech position.

Usually in a *breech birth* the hocks catch on the pelvic bone and the tail is presented through the birth canal. In such cases it is important to work rapidly, because often the navel cord is broken or pinched and the lamb will try to breathe, and drown. Straighten out the rear legs and pull the lamb out as quickly as possible. You can also recognize a breech birth in your initial examination by noting if the hooves are pointed down. (Hooves are pointed up in normal presentation.)

In aiding a ewe, don't panic. Make a complete and careful examination and go through all possibilities in your head beforehand. Don't just grab and pull. Be sure to have all the equipment needed at hand before you begin. And don't turn a lamb to the breeched position except as a last resort. And, above all, don't be afraid to call on someone more knowledgeable than yourself. Let the ewe mother a lamb you've aided normally and make sure you administer a uterine capsule or a penicillin shot after inserting your hand or any foreign objects into a ewe, to prevent infection.

Problems After Lambing

Make sure the newborn lamb begins breathing. Remove any membranes and mucous from the nose and mouth. If this doesn't work, a slap on the rib cage or dropping the lamb on the bedding may do the trick. Blowing in its mouth or artificial respiration can also assist breathing.

If a lamb is quite chilled it should be immersed in a pail of water as hot as one's elbow can bear (90°-100°). Leave it in (head out, obviously) for 10 to 15 minutes, towel it off quickly and place it under a heat lamp. The danger of this method is that much of the lamb's odor will be washed off and the ewe may disown it.

It is important that the lamb nurse within the first hour. The first milk—colostrum—is rich in vitamins and has an important laxative effect (and other properties that are not completely understood). If the lamb does not nurse within a half hour or an hour, again be patient. Allow the lamb to gain its strength slowly. You may have to aid it. Put some milk on the end of the teat and push the lamb's mouth to it or strip some milk into its mouth. Tickling the underside of the lamb's tail while it has the teat in its mouth also helps.

If a mother doesn't clean her lamb off, place it by her head to encourage her. If she still doesn't, dry it yourself with a clean rag or towel. If she won't accept it or let it milk, you will have to aid the lamb. Often the cause of a ewe disowning a lamb is that she is a first-lamber, she is in a rundown condition and instinctively knows nursing will be unhealthy for her, or she has udder problems. In the latter case, pressure and pain in her udder may make it uncomfortable to have her lamb nurse. Withholding feed the day after lambing will go a long way toward preventing such problems. Try to strip some milk out of the ewe to lessen the pressure. Apply hot packs and keep the lamb nursing if she'll let it.

If she still refuses the lamb, then you can only counter with perseverance and patience. Don't give up. On one occasion we worked for five days with a ewe before she finally allowed the lamb to nurse. You must try to get some colostrum into the lamb, either by holding the ewe and letting him nurse or by stripping some from her teat and giving it to him with a dropper. Rubbing some afterbirth or some of the ewe's milk on him will often help her to accept him. A little grain sprinkled on the lamb may make her more receptive. Something to distract and worry her, such as a dog tied within her view, will get her mind off her resistance to nursing. Hold her on her back and let the lamb milk, or face her into a corner so she can't move and hold her back legs so she can't kick him away. *Keep at it.* Remember, each rebuff discourages the lamb even more; however, each successful nursing heightens the lamb's confidence and begins to make the ewe more accepting. I happen to believe if you really keep at it you can make a ewe nurse a bear cub! The thrill of seeing a ewe accept a lamb after five days is well worth the effort. If these measures fail, give her another chance the next year. Should she repeat the performance, cull her.

[*Editors' note.* Sometimes the lamb refuses the mother. The author of this book happened to be away when one of his older ewes lambed. Since one of our own pregnant ewes was being boarded on the author's farm and was due any day, we had stopped by to check and stayed to give the stock and house sitter a hand. The ewe lamb looked fine, but she wouldn't nurse, even though the mother seemed calm and cooperative. We were also puzzled by no signs of the afterbirth, and wondered if a twin was on the way, but after four hours we gave up hope of that. Aware of the infant's immediate need for colostrum, we tried every device we knew and had used before. The infant would suck on a finger but not a teat. The lady editor managed (for the first time) to strip the ewe while one of us held the lamb's mouth open, but that wasn't really very effective. We called the vet to make sure we were proceeding in the right direction. He told us the ewe could have eaten the afterbirth—though we could reach in and find out if we wanted to be sure (we didn't)—and that we should give the lamb a couple of ounces of the ewe's milk every hour or so in a baby's bottle with a nipple. We did so, after borrowing a bottle from a neighbor. *Keep one on hand for just such emergencies.* The nipple hole had to be considerably enlarged, and although the lamb did not swallow much, it was evidently enough to prompt her to start nursing normally as soon as we put her back in the pen with her mother. Lesson: even rank amateurs can rise to the challenge of lambing!]

If a ewe will not milk her lamb, or in the case of an orphaned lamb, you have two options: raise the lamb by yourself or try to "graft" it onto another ewe that has lost her lamb. A situation where grafting may be attempted (i.e., a lamb is orphaned or rejected and another ewe loses her lamb at the same time) will rarely present itself in a small flock. However, the process is worth mention. Graft only in the case of a mother whose lamb has died—do not attempt it on a ewe who has lost one of twins. Dip the lamb in warm water to wash away its odor, dry it off and rub it with the afterbirth of the dead lamb. Let the ewe lick it and clean it off. Skinning the dead lamb and tying the hide on the lamb may also help. Restraining the mother and applying the same methods as with a real mother who rejects her own should produce results. Keep them confined in a lambing pen for a number of days and make sure she lets it milk and doesn't injure it.

In the case of a lamb that must be raised without a mother, foster or otherwise, you can feed it by bottle or use a pail with a nipple on it as you would with calves. I would recommend using the pail for a number of reasons: first, they can milk themselves and you will not spend nearly as much time as with bottle-feeding; secondly, a bottle lamb becomes a cosset, which is the type of lamb that followed Mary to school one day. A cosset becomes as dependent on the milker (you!) as it would its mother, and goes bananas when you leave it. The bleating will drive you crazy and when it comes time to butcher—well, could you really do it?

It is very important to get some colostrum into the lamb as soon as possible after birth. If none is available from the dead or abandoning mother or another ewe, mix a teaspoon of mineral oil and a teaspoon of cod-liver oil into the milk for the first day. This will give the lamb the laxative and vitamin A benefits of colostrum. Regular cow's milk and conventional milk replacers are not ideal for raising lambs because they are too low in fat. A minimum of 30 percent fat and 25 percent milk protein is necessary. LAMA, a Carnation-Albers milk replacer, is specially formulated for lambs. If this is not available, you can add some pig lard or butterfat (do *not* use vegetable oils) to each pail of milk. You may have to use a dropper to feed for the first few days, but get it on the pail as soon as possible.

TABLE 4: Feeding Schedule for Orphaned Lambs

Age	*Amount*	*Frequency*
1-3 days	2-3 tbsp.	every 2 hours
4-5 days	3-4 tbsp.	5 times a day
6-7 days	½ cup	4 times a day
2-3 weeks incl.	¾ cup	4 times a day
3 weeks to weaning (3 months)	until full	3 times a day

It has been found that lambs do better on cold milk than warm because it will not sour so quickly and they will drink it for more frequent and smaller feedings. This is also easier for you. If you are lucky enough to feed it by the pail (and freezing is not a problem), simply let the lamb feed free choice from the pail. Be sure to change the milk daily and wash and disinfect the bucket and nipple. If it won't accept cold milk, warm it at first and gradually decrease the temperature. If the lamb can be put out to pasture soon it will help as will creep-feeding.

Pinning, or the sticking of the young lamb's tail to its anus, can be a problem for the first few days and may even cause death. Check your young lambs by picking up their tails and loosening them, cleaning off the excrement if necessary. When they are a few days old the bowel movement becomes firmer and pinning is no longer a problem.

HEALTH

The chart on page 204 of the *Appendix* shows some of the more common sheep afflictions, their causes and treatments. Again, a good veterinary book as listed in the *Appendix* goes into more detail and is invaluable. Your veterinarian can also advise you.

Stomach worms are the most common affliction of sheep. They are best controlled by the rotation of pastures. If you change the sheeps' pasture every six weeks the worms' life cycle will be broken. We generally worm our sheep every six weeks, or every time we rotate pastures and every six to eight weeks in the winter. While pigs and chickens are subject to worms, they usually do not cause any great harm. But a worm-infested sheep can get severely rundown and may die if left untreated. A commercial lamb wormer should be available at any feed store. Tramisol is a good medicine that kills the widest variety of worms, including lung worms. Thiabendazole is also good, but does not kill lung worms. Substitute it for the Tramisol every third worming, so that an immunity to medication is not built up. In severely infested sheep, where a heavy worming might be too much of a strain, give half the normal dose and follow it in four days to a week with the full dose. If you are not feeding a grain ration at this time, give them up to ¾ quart a day to help build them up. Coughing, sneezing and a runny nose are common symptoms of worms. A bolus gun, available from most feed stores, is the easiest and surest way of administering worm capsules. In the absence of this tool, the pill may be ground up and put in a small portion of grain, but often the full dose is not so consumed. This is obviously a problem with large numbers of sheep. When using the bolus gun, keep the sheep's head parallel to the ground and don't be afraid of inserting the gun far down the throat, or it will be easy for the sheep to expel these large pills. It is easiest to insert the gun in the side of the mouth and then work it down the throat. After administering the dose, hold the mouth closed for a few seconds to insure that the medicine has been swallowed. It helps to have two people for this: one to hold the sheep's mouth open, and the other to get the pill down. Read carefully all directions for worming medicines. Nursing ewes can be safely wormed, but do not use the medicine on meat animals within 30 days of slaughtering.

In a large flock, external parasites (lice and ticks) are controlled by dipping the animals; but with a small flock, drenching the sheep with a solution mixed in a sprinkling can will treat as many as 25 sheep at one time. Treat sheep at least once a year, and more often if new stock is added to the flock, because sheep introduced from the outside are the main source of these pests. Do not apply these solutions in cold weather or chilling may result.

TABLE 5: Sprinkling Can Mixtures for Control of Ticks and Lice

(1) 0.5% ronnel (Korlan) emulsion in water. Mix 8 fluid ounces (1 cup) of 24% Korlan emulsifiable concentrate in 3 gallons (sprinkling can) of water. Use 1 quart per sheep (12 sheep per canful). Ticks alone can be controlled at the reduced rate of 4 fluid ounces/3 gallons of water and only apply 1 pint per sheep.

Precautions: Do not use ronnel (Korlan) in confined spaces. Do not repeat within 2 weeks. Do not market sheep within 28 days of treatment. Do not treat sheep under other stress. READ LABEL.

(2) 0.6% Diazinon suspension in water. Mix ½ ounce 50% Diazinon Wettable Powder in 3 gallons of water. Keep stirred to prevent powder from settling. Apply 1 quart per sheep. Effective only against ticks at 1 pint per sheep.

Precautions: Do not use this preparation if there are lambs less than 2 weeks old in flock. Do not market sheep within 14 days of treatment. READ LABEL.

Maggots will infect any open wound or wet areas caused by urine and can possibly be fatal. Warm weather is the time for maggots. If your sheep twitch a lot or stamp their feet during warm weather, they may be infected. When sheep are put to pasture before shearing, be sure to clip around their dock so as to not allow urine to wet the wool. The best routine is to shear sheep about two weeks before you are going to put them out to pasture. Then by the time they are set out the nicks from shearing will be healed, and you can drench them at the same time. You should check your sheep often; especially tethered ones, to locate any wounds or areas that are infected or ripe for infection. Remember, conventional antiseptics will not kill maggots. A commercial veterinarian-supplied spray is available for this purpose, or you can make your own solution:

9 parts linseed oil
8 parts benzol
2 parts pine tar
1 part carbolic acid

Trimming hooves once a year, usually when they are sheared, is important in good flock management. Set the sheep on its rump with its head resting on your left thigh. This position will normally quiet the sheep down, but do not hold it in this position too long, or breathing may be hampered. Trim the excess hoof off the sides so it is nearly level with the sole. Do not cut too deeply or lameness may result. If the foot is soft, a penknife can be used for trimming and, if it is harder, pruning shears should be used.

BUTCHERING

While the details of slaughtering and butchering deserve a book of their own, remember that sheep are one of the easiest "large" animals to butcher. After watching it done once or twice, and with the help of a good guide (see *Appendix*), most stockowners can easily accomplish it themselves in an hour or less.

A lamb will dress out at about one-half its live weight. A grass lamb, with minor variations due to breeds, will furnish 30 to 40 pounds of meat when butchered in the fall. Lamb, technically, is a sheep less than a year old; a *yearling* is aged one year to 18 months; and *mutton* is any sheep over 18 months of age. The older sheep get (or any animal, for that matter), the stronger the meat tastes; however, while mutton may be a bit coarser, the popular belief that it is no good is a total fallacy as far as I'm concerned. Commercial mutton, maybe. But really home-grown mutton or lamb is another slice of meat altogether.

SHEARING

Shearing should be carried out in the spring after the weather has settled. There should be no danger of severe weather that would chill the sheep, and it is advisable to wait until there has been enough warm days to bring out the oil in the fleeces. This oil strengthens the wool and also makes the wool easier to clip. Try to shear before setting your sheep to pasture, as new pasture tends to loosen the bowels, which results in stained fleeces. Proper tagging will, however, cut down on staining if you must put them out.

You should remove all stained pieces and burrs (if possible) before shearing. You'll see that it's wise to dig up any burdocks in your pasture because a sheep that gets covered with burrs will have a totally worthless fleece.

It is very difficult to describe the process of shearing in words. You would do well to watch someone and then try it yourself. You can either use electric clippers or hand clippers. It is important to remove the fleece in one piece, shearing close to the body but avoiding nicking, and avoiding any second cuts. Do not go over the sheep again, as short fibers result and the value of the fleece is lower.

If you do not wish to shear your sheep yourself, there are often traveling shearers who will do the job for a few dollars. Often they will not come by for one or two sheep; if your flock is small, arrange for neighbors to gather their sheep with yours in one place. The fleece that keeps your sheep warm the previous winter will keep you warm the next winter and for many winters to come.

In most states and areas where sheep are raised, there are associations of breeders which hold informal, periodic meetings. Beginners especially will find these useful—and fun. You can usually watch demonstrations of shearing, docking or castrating, and the almost incredible herding performance of trained sheep dogs, notably the Border Collie. People often bring two or three sheep to be sheared on the spot for a small fee (about $2). Different breeds may be on exhibit; extension service representatives or veterinarians give lectures and hand out helpful informational circulars or bulletins. Frequently, there will also be spinners and weavers at work, people from your locale or region who will buy or trade your fleeces for skeins. These field days are enjoyable opportunities to compare notes and problems with large- and small-scale sheep raisers—or with other backyarders.

CHAPTER FOUR

Goats

The dairy goat is the animal for the family that has reached its limit in paying the climbing prices for milk and other dairy products. A good doe will produce two to six quarts of milk for up to 200 days per year. By having two goats and breeding them at different times, you can have at least one of the goats milking all year, and have some surplus for sale or to feed to other stock.

Those of you who are not familiar with goats and goat's milk should realize that you are in the minority—more people in the world drink goat's milk than cow's milk. Goat's milk is, in my opinion, totally indistinguishable from cow's milk (maybe a bit better, to quote some "unbiased" sources) and even more digestible. Because of the absence of tuberculosis in goat's milk, pasteurization is not necessary. And don't forget other dairy products, cheese and butter, as well as leather from the hides and chevon (goat meat). And all this for a fraction of the cost and space requirements of keeping a cow.

Being ruminants, goats are also ideal for our protein conversion purposes. While they have the same ability as sheep to convert low-protein feed into high-protein products, they are perhaps even more efficient feeders because they are less picky about what they consume. They are truly a browsing animal—like deer—and while they will consume grain and hay, they also delight in shrubs, weeds, saplings, bark, hardhack, and the like. They would rate number one in this book in feed efficiency, and they also rate number one in personality. If you've ever had a goat you know what I mean; and if you're getting one, you'll find out.

BREEDS

There are five recognized breeds of goat that predominate in this country, and you will be considering either a purebred or some cross of one or more of these breeds.

The first three, referred to as the "Swiss Type" because of their similar conformation and origin, are the French Alpine, Saanen, and Toggenburg.

French Alpine: Similar to the other Swiss breeds in conformation, the French Alpine can be almost any color combination. They have prominent eyes, erect ears, straight (as opposed to dished) face, and a graceful appearance.

Saanen: Similar to the Alpine but with a dished face. Color is pure white to cream color. They are the highest milk producers in the group, but their milk has a relatively low fat content, a minus for those who wish to make a lot of cheese and butter.

Toggenburg: The smallest of the Swiss type, with the same basic conformation as the Saanen. They are brown to light fawn in color, with white markings in the form of stripes on the face, white ears with a dark spot in the middle, and white on the legs, near the rump and around the wattles.

The two other major breeds:

American La Mancha: A relatively new breed that gives the distinct impression of having almost no external ears. They have a straight face and various color combinations. They are quite calm and good milkers.

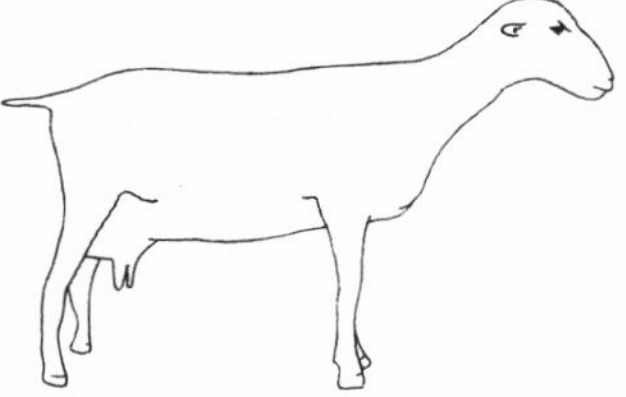

Nubian: A popular breed with long, droopy ears and a roman nose. They have a variety of colors and combinations. They are most noted for the high butterfat content of their milk, but the lowest (on the average) volume of production. Good for making a wide variety of dairy products.

PURCHASE

Within each breed you will come across different types of breeding: purebred, American, recorded grade, crossbred, and unrecorded grade. For your information when purchasing, a *purebred* is a goat whose parents are both registered and of the same breed. An *American* is a goat that is a result of grading up, by breeding three successive generations of a goat to purebreds of one breed. The result after three generations of breeding will be ⅞ of that purebred and be termed an American (e.g., a grade goat bred with a purebred will produce a kid that is half purebred; the next breeding will produce a ¾ purebred; the final a ⅞ purebred or American). The breed American La Mancha came about by breeding a short-eared breed with a purebred stock. A *recorded grade* is a doe (bucks of this type cannot be recorded) with one parent of unknown or mixed breed and the other a registered purebred.

A *crossbreed* is the result of breeding two purebred parents of different breeds; and an *unrecorded grade* is of unknown or unrecorded parentage.

As far as breeds go, there is not enough difference between one or another to single out any as superior. If you want more by-products such as cheese or butter, you might try a Nubian, but any breed should fit your needs. By far your chief concern should be: is the breed of your choice popular in your locale? You may have the handsomest purebred Nubian, but if there are no others around you'll have a problem breeding. Check with any goat clubs in your neighborhood to find out which are the most common breeds in your area.

What type of animal should you buy: purebred, cross, grade, etc.; should you buy a kid, yearling or older milking doe? There are no pat answers to these questions; it will depend on you. A purebred doe will not necessarily produce more than an unrecorded grade, and she will cost more. However, with recorded animals you have records of their parentage and perhaps milking records. With any unrecorded animal you are gambling. You might get some hints as to future production by conformation, but records are the best source.

Purebred goats will have more valuable offspring. You'll always have kids to get rid of and, if they're purebred, the chances are you'll find a better market (both for does and breeding bucks), and they will fetch a higher price.

If you buy a six-week-old kid you may have to wait up to a year before you begin to get milk. A yearling will cost more, but will be ready to give milk. An older doe will cost somewhere in between a kid and a yearling, but will have a less productive life ahead of it, and is more of a gamble for all but the most experienced buyers. All in all, the prices will work out about the same, so suit yourself: milk tomorrow or this time next year.

Unless you have records, telling the age of goats is not infallible, but you can check the teeth the same way you do with sheep. Forget a buck to start with (more about them in the section on *Breeding*). When you go to look for your goat if possible bring an expert with you and do not be in a rush. Shop around. Again, going to shows or fairs and noticing top-rated animals will help you learn good points.

Body

While we look for blocky, heavy conformation in all the meat animals we are describing, dairy animals should have a more angular conformation: thin thighs, prominent hip bones and a comparatively "lean" (but not sickly) appearance. Any thicknesses of the body such as a short neck or any fatty areas are indicators of poor milk producers. A good dairy goat should be rugged-looking with a broad chest, large girth and with front feet set widely apart. The back should be straight (not arched or swayed) and there should be a gradual slope from the hip bones to the tail. A wide barrel and ribcage are desirable as they indicate good food capacity and a potential for large litters. The ribs should be far enough apart to slip a finger in between. The skin should be smooth, thin and supple.

Feet and Legs: The goat should stand well and erect and move about with ease. Check hooves for foot rot and poorly trimmed feet.

Head: The goat should have bright eyes and an alert appearance. Check the ears for any sores or scabs and avoid a doe with any lumps or malformations of the jaw. Check the teeth to help in calculating age and reject a doe with any missing, excessively worn or broken teeth.

Udder: Obviously, the most important part of a dairy animal. Make sure the udder is well formed (see *Fig. 2*) and free from any injuries, scars, extra teats, etc. It should have a lot of capacity and be held high enough to avoid injury from hitting objects. Look for one that is soft with no hard spots in the udder proper or teats (the udder will be firm and hard if the doe has not yet been milked). While the udder will never be perfectly balanced, avoid one that is terribly unbalanced, and

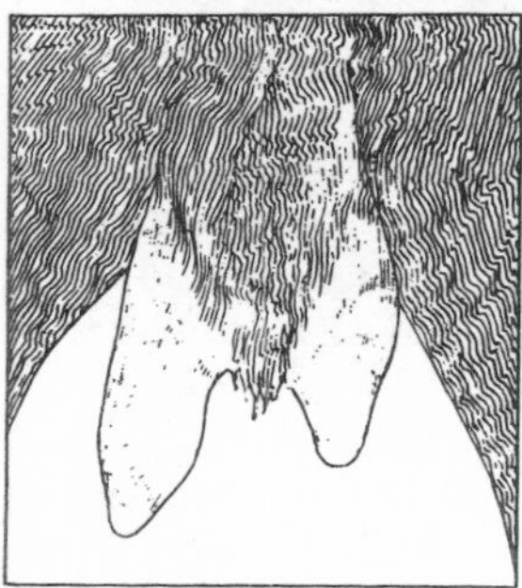

Teats too large

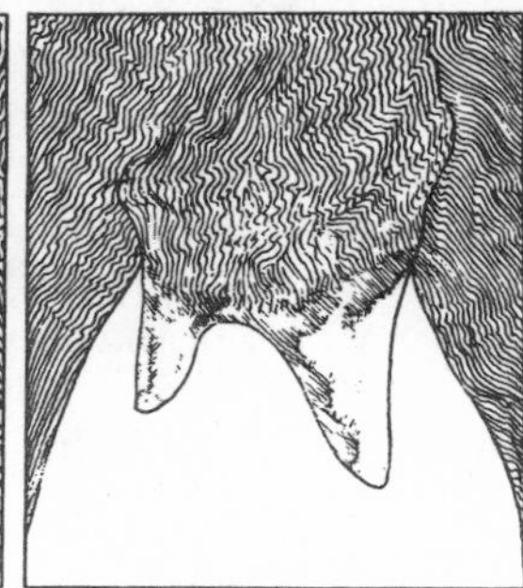

Unbalanced: lacks capacity

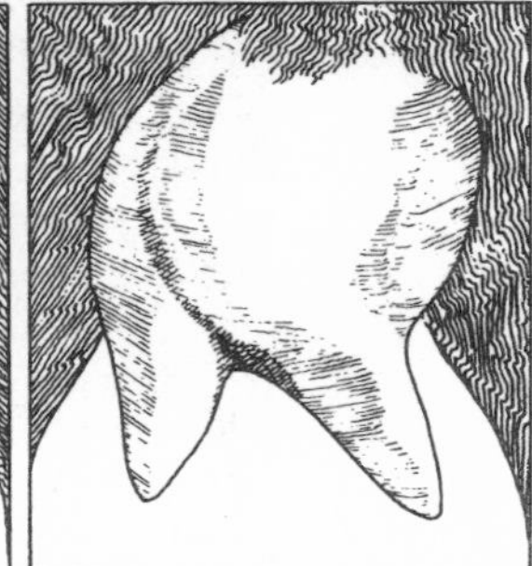

Well-formed

avoid a doe with overly large teats or ones that are pointed sideways. (If they look you in the eye, don't buy.) The ideal teats are the right size for milking and tilted slightly forward.

If you know how, milk the animal yourself to determine capacity and temperament. Check to see if the seller has any milking records of your doe or her ancestors. Records beyond the second generation, no matter how impressive, are of no use and can be misleading. Look for a long milking cycle and high production. Between 3,500 and 4,500 pounds per year is a good average. Records also show that the seller has an interest in his flock and strives to improve it.

HOUSING

Housing for goats is almost identical to that for sheep with a few exceptions. Goats are not as hardy as sheep and need to be more protected from drafts in the winter; and they are better at escaping from fences, so more care is needed there. By reading the section on *Housing* in the chapter on *Sheep*, you will get a good idea of what you'll need, with the following modifications, additions or deletions.

Pasturing

Tethering: Goats, like sheep, can be tethered but they don't take to it so well and extra vigilance is needed. While goats are ruminants like sheep, they browse more like deer. They not only need grasses but a wide variety of feed: plants, small trees, shrubs, hardhack, etc. A tethering system may affect their food intake adversely.

Fencing: Goats are much harder to confine than sheep, and may be one of the most difficult livestock to keep where you want them. Electric fence can be utilized as with sheep, but use three wires, one each at 10, 20 and 40 inches. You'll need a strong charger, and may have to train your goats to adjust to the fence. This, in my opinion, is not the best method.

The USDA recommendation of a five-foot fence for sheep should be adequate for goats. They like to climb on the fence and if it is loose they can, in time, pull it down. Use posts at frequent intervals and get the fence as tight as possible. If this still won't contain them, you may have to resort to the absurd-sounding solution of a fence within a fence. Either use a strand of barbed wire or electric fence (better) ten inches high and ten inches away from the stock wire. A strand of electric wire ten inches from the stock wire and a foot or two from the top may serve to discourage climbing. You may not need all of this, so don't let it scare you off.

Do not enclose any young trees within the fenced area or they will soon be devoured. It is even advisable that such trees be located out of the sight of your goats, lest they become too great a temptation. Goats, like sheep, are sensitive to certain poisonous plants: buttercup, bracken fern, cowslips, false hellebore, dutchman's breeches, water hemlock, mountain laurel, sheep laurel, sneeze weed, white snakeroots, wilted or dry wild cherry leaves, dry oak leaves, rhododendron, hemlock, azalea, milkweed and locoweed, yews, foxglove, delphinium, lobelia and lily-of-the-valley. This is not a complete list; your state extension agent should be able to furnish you with one for other poisonous plants in your region.

Your goats should of course have shady areas. Your gates should be tightly secured because goats are quite adept at getting them open. If you have a buck, he must be kept at least 50 feet away from your does at all times to prevent his odor from tainting the milk.

Drylots: Because fencing a large area for goats can be quite expensive, and because they are browsers and a field may not contain all their food needs, some people may wish to fence a small area, a drylot, and allow the goats to use it for exercise and bring all their food to them. If you do this, you *must* be conscientious about furnishing them with food every day. This is also an option for those people with a very small space for grazing.

Pens

Housing is simple, very much the same as you'd provide for sheep, but be more careful to avoid drafts or dampness since goats are not so hardy as sheep. You can keep your goat (or goats) with your sheep, but be on the lookout for butting, especially if your sheep are a horned breed, which may damage the udder. Allow at least 15 square feet for each goat in your pen. As with sheep, don't have any low doorsills on which a doe may injure her udder. For a larger herd, or in the case of limited space, stanchioning can be used. In this case allow at least an area 2½ feet by 5 feet for each goat.

Unless you have a very large flock you won't want to go hog-wild in construction, so make use of what you have for your goat pens: a part of your barn, an old shed, a garage. The best type of flooring, for sanitary reasons and ease of cleaning, is concrete. You will, however, need a lot of bedding on this to keep the animals warm in winter. Wooden flooring is fine as long as you provide enough bedding to absorb the urine before it soaks into the wood. Dirt and gravel floors, with adequate bedding, are suitable and afford excellent drainage. Any of the most common and available bedding materials can be used.

Milking Room

For goat's milk that is to be marketed commercially, there are many regulations governing the location, size and fittings. Some call for an adjacent room for washing and sterilizing milking equipment, a place for cooling and storing milk, and a place for storing milking equipment. For the small operation this is both unnecessary and prohibitively expensive. You will, however, need to take certain steps to insure the quality of your milk. As you probably know from experience with ordinary milk in a refrigerator, milk collects odors from its surroundings. For this reason, you should do your milking in a room or building away from odors, and preferably free from dust, dirt, cobwebs, flies, etc. You need not supply storage space for milking equipment, since these can just as easily be cleaned and stored in your house.

EQUIPMENT

You will need watering and feeding equipment, milking utensils and perhaps a milking stand.

Hay/grain

Goats are more picky than most animals in that they won't eat off the ground and often won't eat hay that has fallen from a rack onto the ground. If you are feeding them with sheep, the sheep *might* clean up after the goats. A separate grain feeder is not needed because it's a good idea to feed your doe the grain ration when you're milking her. This will make her more agreeable when milking and enable you to keep an eye on her feed intake. A hay rack of the kind used with sheep is satisfactory, as goats usually will not pull out and waste much hay in this type of feeder.

Waterers

Goats naturally need a lot of water to make milk, but they will not touch any water that is contaminated or even dirty. Buckets or troughs should be secured to prevent tipping and placed off the floor a bit so droppings will not fall in them. Be sure to keep water thawed at all times in the winter, and if the tap water is very cold, add some warmer water to it so the goats are more likely to drink.

Milking Equipment

The following list includes milking equipment that is both necessary and optional as indicated:

Small pail with cover (stainless steel is ideal)
Pail and cloth for washing udder
Milk strainer
Storage bottles for milk
Milk stand (optional, see below)
Scale for weighing milk (optional)

Stainless steel pails with half-moon covers are worth the money and very easy to clean. Storage bottles that are free and completely adequate are half-gallon fruit-juice bottles or the like. While you can milk your goat on the ground while it is stanchioned, a milking stand (*Fig. 3*) can be made of wood as indicated or with other random material. Note the place for

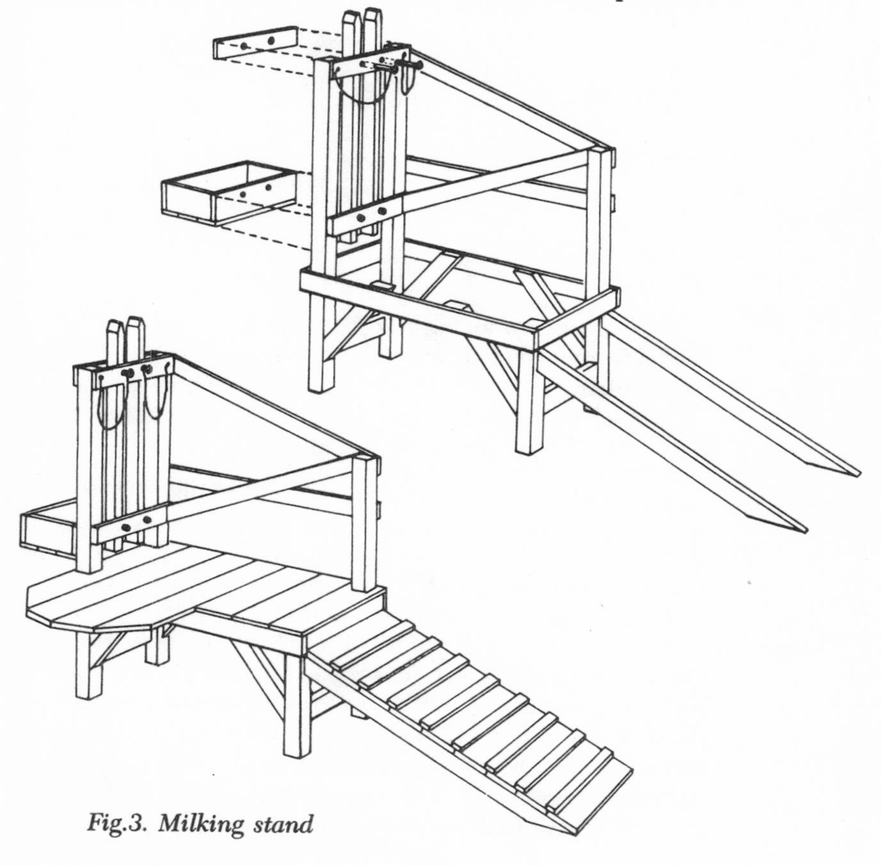

Fig.3. Milking stand

the feed dish. If space is a problem, or if you have a permanent milking room, you may want to construct a foldaway milking stand (*Fig. 4*).

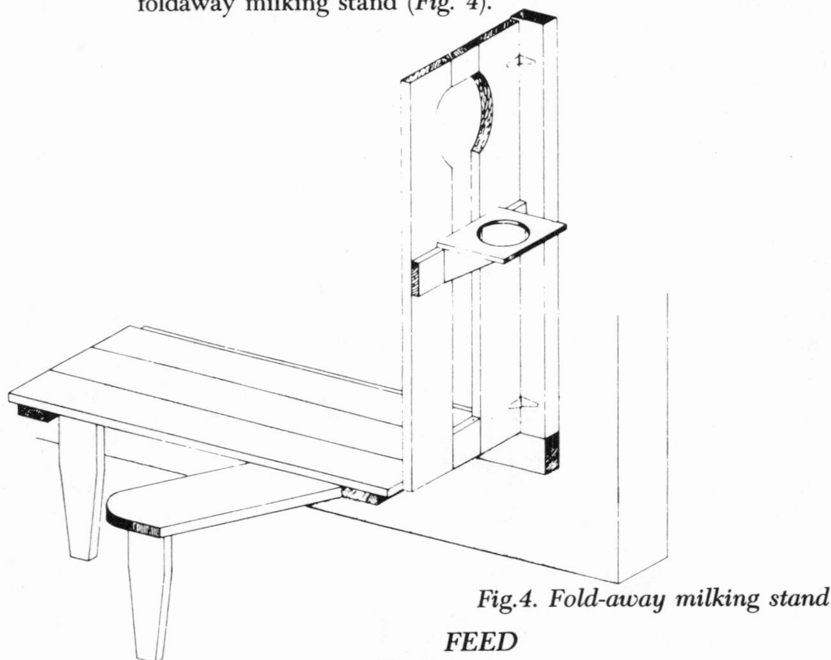

Fig.4. Fold-away milking stand

FEED

While goats are among the best browsers, and will clear hardhack and other undesirable growth from pasture, don't expect them to thrive and produce milk on browsing alone. They need good pasture or hay and a grain concentrate to produce large amounts of high-quality milk.

Pasture

In summer months this will constitute the bulk of their feed. A rule of thumb is ½ acre per goat, but this is meaningless without gauging the quality of the pasture. Use this guideline as a starting point. Watch the pasture to keep it from getting grazed too close and watch the condition of your goats. There can be plenty of weeds and hardhack, but there must also be a high percentage of good grasses and legumes.

Goats do not like (nor will they thrive on) lush grasses alone. Put them on a lush, green suburban lawn and they will be miserable. If you do not have enough browse in your pasture, supply your goats daily with twigs, saplings and tree prunings (especially from fruit trees). Be sure you do not feed, or have in the pasture any plants that are poisonous. If you raise your goats on drylot, you must supply them with as much green and browse as they will eat. If your pasture becomes overgrazed, or in dry times (when pasture wilts and drys up), or toward the end of the grazing season, you will want to supplement their pasture with good hay.

Hay

During winter months hay will constitute the bulk of your goat's diet. The better the hay (i.e., fine, legume hay), the lower your outlay for grain will be (they will need less of it and grain of a lower protein percentage). Hay that is at least half legume and preferably second cutting is preferred. As with sheep, second-cutting hay is worth the extra five dollars or so a ton as there are fewer coarse stems and hence less wastage. If no legume hay is available, any good-quality carbonaceous hay is suitable (see *Chapter Seven,* "Grow Your Own . . ."). Hay should be fed free choice, but when stocking up for winter figure 2 to 3 pounds per day per goat, or about ¼ ton per goat.

Grain

This is the most expensive component of your feeding program and real savings can be made if you can grow some of your own crops to help make your own feed. If you are feeding good legume hay (at least 50 percent legume) or good legume pasture, then the protein content of your grain concentrate need only be 12 to 14 percent. If your feed is grassier hays and pasture, you will need grain with 16 to 18 percent protein. (Feeding of bucks, pregnant does and kids will be discussed in the *Breeding* section of this chapter.)

Yearlings (after weaning until beginning of milk production): You will want to feed growing goats enough to maintain them and for proper growth but not so much as to make them fat. You should be able to feel the ribs of a growing goat. They need plenty of fresh air and exercise. Toys, such as a tire hanging from a tree limb, old barrels or the like will furnish hours of enjoyment (and exercise) for your goat. Depending upon the quality of your hay or pasture (see above) a 12 to 14 percent or 16 to 18 percent protein ration will be needed. A pound to a pound and a half should be adequate, but keep an eye on the animal: check ribs and general appearance in order to lessen or increase grain intake. Of course, plenty of fresh water should be available at all times and a salt mixture of ½ mineralized salt and ½ dicalcium phosphate should be provided free choice.

Dry Does (not pregnant): Dry does can be maintained on good pasture or hay, with root crops (beets, turnips, carrots) and cabbage-family plants thrown in along with twigs from fruit trees. If the doe appears in poor condition, worming may be needed or some grain may be added to the feed program.
Milking Does: Along with the pasture and/or hay, a milking doe will need about ¼ ton of grain per year. This may be commercial grain (choose a grain that is palatable to your doe and of the proper protein percentage) or one of the mixtures listed below:

TABLE 1: Grain Mixtures for Goats

Mixture	Ingredients
#1	Oats 60% Wheat bran 30% Soybean or linseed oil meal 10%
#2	Corn 35% Oats 35% Molasses 15% 30% Protein supplement 15%
#3	Corn 45% Oats 20% Wheat bran 20% Soybean or cottonseed oil meal 15%
#4	Barley 50% Oats 25% Wheat bran 10% Soybean oil or cottonseed oil meal 10%
#5	Oats 100%
#6	Oats 50% Corn 50%

Any of the above grain mixtures can be used interchangeably and to replace commercial grain. As with all stock, do not change the feeding regimen too often or too quickly. If you must change the ration, do so very gradually. The mineral mix mentioned earlier should always be available and, naturally, plenty of fresh water.

For milking does, feed one pound of grain for each three pounds of milk produced (approximately 1½ quarts). One method of feeding is to keep increasing the grain slowly until peak milk production is reached. Then cut the grain back (again, slowly) as far as possible without lowering milk production.

In the summer when goats are on pasture, you need only furnish the grain in the amounts prescribed above; in the winter supply hay free choice and the prescribed amount of grain mixture.

Goats will do better on a number of smaller feedings per day. You should, however, feed at least twice. The grain ration can conveniently be fed twice a day at milking time. This will give the doe something to keep her mind on and make getting her into the milking apparatus less of a chore. Also, if she is housed with sheep, this will insure that she gets her full grain ration and is not being bullied by the sheep.

Supplementing Commercial Feed

Any of the grain mixes previously listed will lend themselves to at-home mixing using home-grown grains. In addition, silages, root crops and cabbages can be fed to goats at a rate of two to three pounds per day which will replace one pound of hay.

John N. Smith of Las Cruces, New Mexico, doing Master's research, has developed a feeding program for goat kids on ensiled tomato vines. The tomato vines were the by-product of a hydroponic greenhouse operation, but conventionally grown tomato plants could also be utilized. The mixture consisted of roots, vines and some immature fruit. It was chopped and treated with acetic and propronic acid (mixed together at a 4:1 ratio) at a rate of 1 oz. per 9 lbs. of mixture and ensiled in small drums (see *Chapter Seven*, "Grow Your Own . . ." on making silage). He fed it from the age of eight weeks on at a rate of ¾ lb. per day per kid. The silage, as fed, consisted of the following:

Dry Matter	15.7%
Crude Protein	2.2%
Fiber	4.7%

In a feed trial comparing the tomato vine ensilage to milo ensilage as a control, the kids on the tomato vine ensilage had a higher average daily gain and a lower feed conversion ratio. Later both groups of kids were offered both ensilages, and both groups preferred the tomato vine ensilage. Contrary to earlier reports of toxicity, no ill effects were found that could be related to the ensilage.

MANAGEMENT

Milking

Hold it a minute. You don't just plop the goat on the milk stand (if you use one), begin milking away, and then guzzle down the product. You must first clean and disinfect all your equipment, wash the doe's udder, then milk, strain the milk, clean your utensils and finally cool the milk.

Equipment Preparation: After each milking all your milking utensils must be thoroughly cleaned. First rinse your utensils in cool water and an alkaline detergent (do *not* use conventional dishwashing detergents as they will leave residues and possibly affect the taste of the milk); place the utensils in the solution and let them soak for five minutes. Once or twice a week use an acid detergent since it will remove any mineral deposits that tend to build up with the use of an alkaline detergent. After soaking the utensils, scrub them with a brush. You will see why good, seamless, stainless steel buckets are easiest to keep clean. After scrubbing, rinse utensils in hot water, invert them and allow to air dry.

Before your next milking, the utensil must be sanitized. Purchase a sanitizing chemical (chlorine, iodine and ammonia compounds are most common) and follow directions that come with it. Usually, they should be soaked in hot water containing the sanitizing agent for five minutes.

General Preparation: Before we get down to the actual milking, a few more preparations should be made. The area around the udder and rear of the goat should be trimmed periodically so long or stray hairs do not fall into and contaminate the milk. Likewise, brushing your doe before milking will remove loose hairs and dirt, and will make her eternally grateful to you. You are what you eat, and a doe's milk is what she eats. Therefore, a meal complemented by milk from a goat that has recently dined on garlic bread and anchovy pizzas would be less than perfect. If you are feeding cabbage or silage to your does, do so immediately *after* milking so that the flavor is not passed into the milk. Feeding such items immediately before milking may taint the milk.

Immediately prior to milking, wash the udder and your hands with lukewarm water containing a sanitizing solution, then dry. Use paper towels for these operations since they are less apt to be contaminated than sponges or cloths you use over and over.

If you're one of those people who have trouble tapping your fingers in sequence when you're impatient, then you may have trouble milking at first. But with practice, anyone can become a good milker. Be prepared, however, to annoy your doe the first few times you try it, and also to get more milk on the stand (and on yourself!) than in the bucket. Watching an experienced milker will help you learn, but if one is not available, follow these simple directions.

First, be firm but gentle in your milking. Patting your doe before you start and talking to her will help you both along. Grasp the top of one teat, as close to the udder as possible, between the thumb and forefinger and squeeze. This will trap the milk in the teat and prevent it from being "milked back" into the udder, Next, while pulling down on the teat gently, encircle the teat and squeeze with the middle finger, next the third finger and finally the pinky. Done in rapid sequence, this is all there is to milking. Then do the same with the other teat, alternating teats until the udder is dry. (The first few streams of milk from each teat should be milked into a separate cup to look for "off" milk that may indicate disease.) As the milk flow lessens, pushing up on the teat and massaging the udder gently will help get the last of the milk out. After finishing milking, strip each teat (similar to stripping milk into the mouth of a young lamb) by grasping the top of the teat and running the thumb and forefinger down the full length of it.

Goats should be milked twice a day (extra-heavy producers may need a third milking) as near to 12 hours apart as possible. Young does may resist milking at first, but by being gentle and feeding them grain they should soon settle down. Also, a full udder is not comfortable, and the doe will soon learn that your milking will give her great relief. Above all, be consistent in your milking: try to milk and feed at the same time each day; always milk from the same side and if at all possible have the same person do the milking; treat your goat kindly and act the same way every day—leave your bad moods outside the milk room. Taking your frustrations out on your doe's udder will not be an aid to good production.

Caring for the Milk: Immediately after milking strain the milk into a storage can or other container. Remember that milk absorbs any odors very quickly, and that it must be chilled to 38°-40° as quickly as possible (ideally, within one hour) for best quality. Chilling cannot be accomplished fast enough in a refrigerator; submerging the milk container in cold water or ice water should do the trick. Change the water frequently since it warms, and when properly chilled, refrigerate. A cool stream would do the job quickly and save the trouble of procuring ice cubes and changing the water, but do be sure the milk is not swept away down the stream to a hot and thirsty fisherman who would be all too eager to accept his catch.

To pasteurize or not to pasteurize? If you plan to sell your milk commercially, you will have to pasteurize. For home use or sale to consenting neighbors . . . well, it's up to you. Pasteurization, or the heating of the milk to kill disease-producing bacteria, if nothing else destroys some nutrients in the milk. Jerome Belanger, a leading authority on dairy goats, maintains that tuberculosis and brucellosis are not problems with dairy goats, so pasteurization is not necessary as it is for cows which are prone to these diseases. This is not to suggest that you can milk into a rusty pail and not get sick from your milk. There are plenty of other bacteria that can get into milk *if* you do not follow the recommended sanitary precautions. In my opinion, you don't have to pasteurize if you maintain sanitary conditions. But suit yourself. Here is a simple, at-home pasteurization method (small, family-sized pasteurizers are also available from Sears and other farm and dairy supply outlets):

(1) Place up to six quarts of milk in a glass or stainless steel kettle or flat-bottom pan. (Do not use copper, iron, or chipped enamel utensils. Copper or iron utensils may cause an off-flavor in the milk.)

(2) Place a floating dairy thermometer in the milk. (Dairy thermometers can usually be obtained in hardware stores or dairy supply stores. Do not use candy thermometers, since these frequently have metal parts which may impart an off-flavor to the milk.)

(3) Heat the milk rapidly, stirring constantly with a stainless steel spoon, until a temperature of 165°F. is reached.

(4) Hold at 165°F. for 20 seconds, then place the kettle or pan immediately into a large pan of cold water and, with constant stirring, reduce the temperature quickly to 60°F.

(5) Store the milk, well covered, in clean containers in the refrigerator at 40°F.

Goats are in their prime from four to six years of age, but with proper care some does will milk satisfactorily to the age of 10 to 12.

Dehorning

While some goats are born hornless, others are not, and dehorning is a good idea to prevent possible future injury to the rest of your herd, other animals, or that favorite animal of all, yourself. Ideally, dehorning (or disbudding) should be done at from three days to a week after birth; much older and you're liable to run into problems or have to turn the job over to a veterinarian. A goat that will have horns shows a swirl of fur on its forehead; a naturally polled kid will not show any peculiarities in the fur. As a further test, check the skin on the forehead. If it is loose and moves, then the kid will probably not grow horns (despite, even, the presence of swirled hair); tight skin on the forehead indicates a horn under there somewhere; and you'd better get it soon. *Note:* Nubians often exhibit signs of horns at a later age, so be careful.

The two most common methods of disbudding are caustic paste and the use of a hot iron. The use of caustic paste is the easiest, but takes a couple of days and in that time it's not unusual for the paste to run and damage surrounding tissue, or even get into the eyes and cause blindness. I do not recommend using it, but if you insist, first clip the hair around the buttons and surround them with Vaseline to protect the skin. Secure the kid in a stanchion or in a disbudding box and *carefully* apply the paste. After an hour let the kid go, but first cover the paste with some Band-Aids. Be sure he can't rub on other goats; keep him out of rain, since water will make the paste run into his eyes; and watch closely to see if he rubs and spreads the paste. (See why this isn't such a good method?) With a hot iron, you burn the tissue around each bud for 10 to 15 seconds, effectively stopping the growth of the horns. The box shown below is invaluable in restraining a kid. It is best to

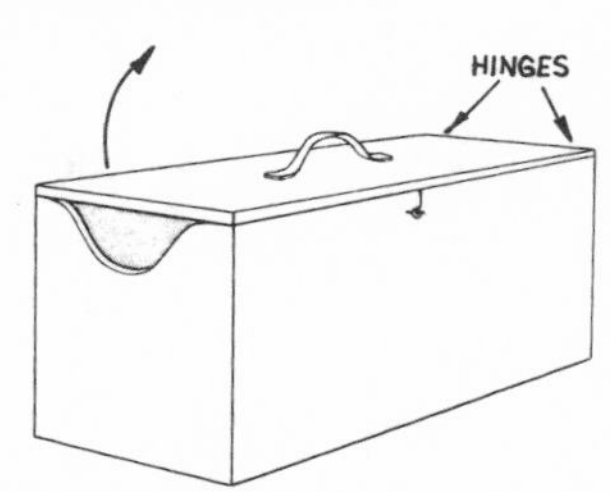

Disbudding box

have this procedure shown to you by an experienced person before you try it yourself. Despite the anguished cries of the kids, take comfort that it hurts the little tyke no more than it would you if an anvil were dropped on your foot. Seriously, once it recovers from the humiliation of being stuffed in the box, it will bounce around as if nothing had happened.

BREEDING

While you can get eggs from a hen without the service of a rooster, you cannot get milk from a goat without first breeding her. Does can be bred at about seven months when they weigh 80 to 90 pounds. After kidding, the kid(s) will be taken away in a few days and the goat can be milked for about 200 days. The doe should rest for two months and should have been bred to freshen at the end of this two-month period, and your cycle begins again.

Bucks

In a herd of the size most people will want—one doe or two at the most—keeping a buck is not necessary. It will cost you far more than the usual stud fee ($10-$20) to keep him for the year, so there are no economic reasons for doing so. Bucks have a rather unpleasant smell, which can taint the milk or your relationship with your neighbors. If your space and olfactory senses allow, you may consider keeping a goat for *convenience* (this is especially true if you have several does). Like sheep, does do not always exhibit signs that are discernible to the human eye or nose when they are in heat. A number of false alarms and wasted trips to a breeding buck may convince you to buy your own buck. If you import your hired sire and have trouble discerning heat, you'll have to have him (and his milk-tainting odor) around for a while, and expose him to the doe once a day to see if she'll accept him. This, too, can be a nuisance. In any event, try hiring a buck the first few times and see if it works. If you do decide to buy a buck, look for the qualities mentioned in selecting a milk goat. Be sure that he has two testicles that are fully descended. A good-grade buck can do quite well, but if you have purebred does you might consider a purebred buck of the same breed, since any kids can be sold for a higher price.

Feed enough to maintain your buck and don't let him get overly fat, since that may make him lazy in his breeding. Good pasture can maintain an inactive buck, as will hay and a pound of grain in winter months. Beginning two weeks before, and during the breeding season, allow one to two quarts of feed in addition to his normal ration—more, if heavy demands are being made on him and he looks out of condition. Again, do not allow him to become overweight. While a buck is potent at about four months of age, don't use him for breeding until he is seven months old. He shouldn't be used more than once every two weeks at this early age, but can handle four or five does in his second year. Goats are hard to pen, and bucks that want to breed are even harder. Be sure to securely pen your buck away from your does when you don't want his services.

Does

As mentioned above, does can be bred at seven months, but definitely not before they reach the weight of 80 to 90 pounds. Artificial insemination is becoming more widespread for goats, and you might want to check with your extension agent or local goat clubs to see if it is available in your area. Like sheep, goats are seasonal breeders. They can be bred from late August to late February, or as late as mid-March. They come into heat every 18 to 24 days (average: 21) for a period of 1 or 2 days during which time they'll accept the services of a buck. The gestation period is from 145 to 155 days (average: 150). The following chart will help you determine when your doe will kid:

TABLE 2: Gestation for Goats
(Based on average gestation period of 150 days)

When Bred In:	*Will Freshen:* (Breeding day less number below) *	
	Month	Day
July	December	–3
August	January	–3
September	February	–3
October	March	–1
November	April	–1
December	May	–1
January	June	–1
February	July	0
March	August	–3
April	September	–3
May	October	–3
June	November	–3

* To determine day due to freshen take breeding day and subtract the number indicated on the table. For example, if bred July 10 doe would be due to freshen December 7; if bred November 20 she would be due April 19.

While you should rest your goats (i.e., dry them up) two to three months a year, you can breed them in such a way as to never be without milk—if you have more than one doe. And at the time when they're both producing, you'll have surplus to sell or feed to other stock.

Detecting heat in does is not as easy as it may be with sows. They will exhibit much the same signs as sheep with the additional helpful hint that their milk production may fall suddenly. Coupled with the other signs, you should be able to learn to tell the heat period with some accuracy. Remember that young bucks can impregnate does at four months of age, so separate them from does early on, or castrate ones not wanted for breeding.

You can figure the approximate kidding time by consulting *Table 2.* If your doe is milking, you will want to dry her up about two months before she is due to kid. There are two reasons for this: first and foremost, she needs this two-month rest each year; second, most of the fetus' growth occurs during this time and the strain of producing milk and contributing to the growth of a kid would be considerable. Drying up can be accomplished by cutting the grain ration and simply not milking the doe. After a few days of discomfort, the milk will be absorbed back into the body and she will be dried up. Plenty of roughages will also aid her in this process. After she dries up you can gradually increase her grain to a point where the doe is in good shape but not too fat. First-kidders which were not milking when bred need feed for their own growth as well as that of their kids. Feed to achieve the same body state as with older does; start with one pound per day up to six weeks prior to kidding and 1½ pounds from that time to kidding.

Kidding

Kidding is almost identical to lambing, including the cutting back of bulky feeds a few days before kidding, to the actual birth process and possible problems. The chapter on *Sheep* goes into considerable detail in these matters and can be used as a guideline in kidding. The major exception is that multiple births are much more common in goats. Twins after the first year are very common, and triplets and quadruplets are not uncommon.

Feeding Kids: As with all stock, make sure the young kid(s) get the mother's colostrum. If you are going to let the mother rear the kids, then let nature take its course. However, since you have your goats for milk it is unlikely you'll want to give up your milk for the three to four months it will take before the kids are weaned. Try not to let the kids nurse if they are to be raised artificially. First, milk some colostrum out of the mother and feed it to each kid, by bottle or eyedropper if necessary. Thereafter, milk the does as you usually would, but you will not want to use the milk for a few days to a week until it reaches its normal quality (feed it to the pigs or chickens in the meantime). It will take her about a month to reach the peak of her production again.

You can raise the kids on bottles or from pans, but the former method is preferred since less digestive upsets from the ingestion of air will result. You can use cow's milk or a milk replacer powder (as with a veal calf) but be consistent in whatever you feed. Warm the milk 103°-105° F. and feed as much as they will consume in a ten-minute feeding three to five times a day. Offer grain and good second-cutting hay (if possible) at the start of the second week. Water should be provided at all times. When you wean the kids from the milk depends on their intake of hay and grain and the quality of any available pasture. It is absolutely essential that all feeding equipment be sanitized before each use to prevent disease. A rack holding bottles can be utilized for self-feeding.

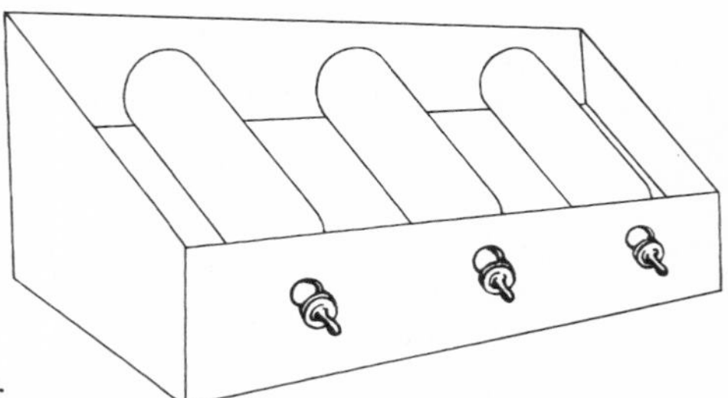

A self-feeding bottle rack

For your own convenience, try to sell doe kids as soon as possible after birth. Buck kids, unless you have purebred parentage, will not be in demand for breeding and should be castrated. This can be done within the first week or two, and the procedures are identical as those for sheep. Tattooing of ears is required for purebred stock, and you will have to buy or borrow a tool for it if you raise purebreds. You will want to identify non-purebred kids in a litter, but rather than buying a tattooing tool, use an ear notcher or put numbered tags in their ears. In smaller litters, or in small flocks, goats can be identified by markings or by different colors.

HEALTH

Mastitis, or inflammation of the udder, is perhaps the most serious affliction that can affect a milking goat. It is caused by bacteria that enter the udder from cuts, scratches or bruises. Symptoms include no milk and an odd walk. The udder is usually quite hard and often hot. Treatment consists of hot packs applied five times a day, following by drying of the udder and milking out. Drugs may be prescribed by a veterinarian, but be sure they are out of the doe's system before the milk is used for *any* purpose.

Internal parasites, particularly ascarids (roundworms) and lungworms are a problem just as they are for sheep. Symptoms are the same, as is treatment. Thiabendazole is effective for roundworms, as in sheep, but does not kill lungworms. Tramisol will kill the lungworms, but care should be taken when feeding it to very rundown goats or pregnant ones, since it is much stronger. Milk taken within 96 hours after worming must not be used for food. Do not treat goats within 30 days of slaughter for meat.

Common goat diseases are listed on page 214 of the *Appendix.*

BUTCHERING

Chevon, or goat meat, can be had by butchering anything from a newly born kid (if you can do it) and cutting it up almost like poultry, to butchering an old cull doe and using her for sausage or stewing. Of course, you can butcher at any stage in between. The meat is quite good and tastes like, and can be cooked like, lamb. Techniques of slaughter and butchering are similar to those used for sheep.

CHAPTER FIVE

Pigs

Pigs have long been the victims of a false image: they are reputed to be unclean, unfriendly and uncouth; they are the unwilling villains of George Orwell's anti-utopian novel, *Animal Farm*; and their name is reserved for the most despicable of people. Happily for us, none of these rumors are true. Years ago a farmer would raise two, keeping one for his family and selling the other. The income from the sold hog would pay for the other and leave extra money for "payin' off the farm." Hence the more appreciative name, "mortgage lifters."

Given the chance (that is, not being confined to a pen the size of a linen closet), pigs will be the cleanest of all farm animals. They are, more importantly, one of the best converters of feed to meat and can make use of pasturing, table scraps and garden surpluses to reduce feed costs. In only five or six months they will provide an average of 150 pounds of pork products. They also require relatively little space and care and are the smartest of farm animals. But *beware*—they are also very easy to make into pets, a decidedly negative factor if you plan to eat them. They will stand by your side for hours emitting squeals and love grunts as you scratch them. They are truly the sweethearts of the barnyard. Much of the violence and hatred of the sixties between students and police could have been avoided if others had agreed with me that the epithet "PIG! !" is the highest of compliments.

BREEDS

With pigs, as with most of the animals we are discussing, there is no one outstanding breed. Your choice will be based on personal preference and on what is available in your locale. Rather than trying to choose any one breed, you should concentrate on finding outstanding specimens within a given breed. While certain breeds may have their own advantages and disadvantages, any well-bred pig will be suitable for home consumption.

Pigs are categorized into two broad groups, lean- or meat-type and fat-type. Years ago, when such "fatty" byproducts as salt pork and lard were more in demand, the so-called "lard" pig was most common. In recent years, through selective breeding, pigs have lost the excess fat and most of them are the leaner meat-type. You should avoid the fat-type. Meat-type pigs are long and trim with hams and shoulders that are wider than their backs. These pigs will dress out with a lower percentage of fat and for that reason are cheaper to raise, because less food goes into the production of fat than meat. They will still, however, provide plenty of salt pork and lard for the average family.

Fat-type

Meat-type

The most common breeds of swine are:

American Landrace: The American version of the Danish hog that has made the Danes famous for their fine hams and bacon. They are white or pink, with floppy ears and a long, lean body.

Berkshire: A medium-sized hog that is very solid and has little excess fat. They are black with white on their feet and often a

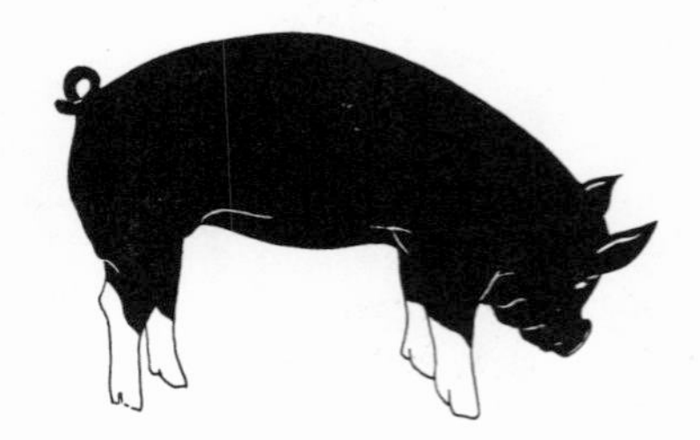

bit of white on their faces. They have broad faces that are slightly dished out and have medium-length snouts. While this gives them a slightly pug head, those that are extremely pug-headed should be avoided.

Chester White: A large, white hog with a medium snout and floppy ears.

Duroc: A lean, hardy hog with large, floppy ears. They are red without a trace of any other color. The color may range from deep rust to almost tan. They tend to have relatively large litters.

Hampshire: This hog is quite popular in the Northeast and is quite easy to care for; it is a good pig to begin with. It is black with erect ears and a white band around the front of the body and forelegs and, sometimes, white on the rear feet. It is a good hog for southern climates because its darker coloration makes it less prone to sunburn.

Poland-China: This hog is black with white feet, and splashes of white on the face and the tip of the tail. It is also particularly suited to warmer climates because of its darker coloration.

Spot: It resembles the Poland-China in body type but has more white. It tends to look more like a black hog with numerous white spots.

Tamworth: One of the oldest breeds of hog, the Tamworth, like the Duroc, is red, varying from light to dark, but with erect ears and a rather long, thin snout.

Yorkshire: This is quite a popular breed and also easy to keep. They are pink or white, sometimes with black spots, and have erect ears. This is the best breed, in my opinion, if you plan to go into breeding, because they have large litters and are very good mothers.

It is unlikely you will be offered, or would want to buy, a purebred pig. A purebred will cost a good deal more, and unless you're planning to go into breeding you'll be wasting your money. I have had personal experience with Hampshire-Yorkshire crosses and Yorkshire-Duroc crosses and have found them quite acceptable. In fact, pigs, like lambs, often benefit from crossbreeding by having increased vigor and growth capability (heterosis).

PURCHASE

A word first about when to buy. Pigs are usually farrowed in the spring and early fall. To start, I would suggest buying a spring shoat. This pig will be raised over the summer months and butchered in the fall so there will be no need for winter housing; moreover, more table scraps and garden surplus are available in the summer, and it is easier to learn about the care for an animal in the warmer months. Shoats are more expensive in the spring, but it is worth the extra money to raise them in the summer. It is also best to raise more than one pig at a time. If your family wants only one, see if a friend will buy another and raise it with yours. Any animal does better with companionship, and if you ever watch two pigs you will notice that they never eat alone. If one gets up to eat, you can bet the other will not be far behind. In a way, they are competing for the food, eat more, and tend to gain faster.

Fortunately for mailmen, hog breeders do not have a mail-order service for pigs as poultrymen do for chicks. This understandably limits your choice of breeds and means a bit of transportation on your part. Your best source is a neighbor who is raising pigs or someone you are referred to by a satisfied customer; otherwise keep an eye out for advertisements in market bulletins and newspapers or word of mouth recommendations. You might also try a livestock auction if there is one in your area, but be sure you know your pigs or have a knowledgeable person with you. Personally, when I buy shoats I never inquire about breed. Again, unless you plan to go into breeding, a meat-type pig of any breed or crossbreed will do well with proper care. In most areas there are never enough pigs to go around, so be happy to get a healthy one, whatever the breed.

The best way to transport your little piglet to its new home is in a small wooden box with some food inside it. Most pigs travel quite well (and do almost anything else well for that matter) if well fed. If you have access to a dog travel box, this will be quite satisfactory; or you can build a simple substitute. If nothing else, they can be transported for *short* distances in a burlap sack. Do not make the mistake of thinking you can hold your cute little piglet in your lap on the way home like a new puppy. It would probably sour you on pigs forever.

Pigs are weaned from about four weeks of age and sold at six to eight weeks of age. They should weigh from 20 to 30 pounds at six weeks and from 30 to 40 pounds at eight weeks. Obviously, lighter pigs may be runts or less than six weeks of age and should be avoided.

Check the going price for pigs in your area, because it is not beneath some people to sell them to you for far above the going rate if they think you'll pay it. If you can't determine the going rate, a general formula for the price of shoats is 1.5 x Weight of Shoat x Average Price of Dressed Pork.

Piglets at six weeks of age have already been weaned from their mother and are eating dry food on their own. Iron shots, which help prevent anemia, should have been administered to the piglets at birth. It is not essential, as some people say, that piglets have their needle or baby teeth clipped. This is usually done in larger litters where competition for food may cause the piglets or the sow's udder to be injured. I never clip our piglets' teeth unless they are causing damage to the udder or themselves, because I feel their removal may hamper the ingestion of solid food early in life, and might invite infection.

There are no advantages, such as speed of growth or meat quality, in either a male or female pig. (Although some say that males eat a lot more and consequently gain more.) Any male should, however, be castrated or the meat will have an undesirable taint. You can tell a barrow (castrated male) by one or two scars near the hindquarters from castration. If the scar is well healed it may be necessary to feel around the scrotum to be sure there are no testicles present. To determine sex of the shoat, check the hindquarters, as the female (gilt) has a small flap and the male does not. Also, in the male the sheath which houses the penis is quite apparent on his underside.

In general, choose the largest and most active pig in a litter. The belief that a pig with a curly tail is better than one with a straight tail is, in my opinion, a fallacy. Look for qualities embodied in a good meat-type hog: long, lean, with large hips and shoulders. Arrange to buy your pigs early in the season and take pains to pick your pig up early (even go a day earlier than suggested; they'll sell it to you anyway and you'll probably get the pick of the litter), or you're liable to get stuck with rejects. Any pig that looks listless, sickly, coughs or shows any other signs of disease should be avoided. This is also true of malformed pigs or those that are humped up or whose vertebrae you can feel. A lump or bulge near the hindquarters may indicate a hernia. Although hernias often heal on their own, choose another to be sure. If purchasing a barrow that has recently been castrated, make sure there is no inflammation or any sign of infection around the incision. Lastly, if buying from a stranger be wary of any pig that looks different from the rest. People have been known to pass a runt on from litter to litter. That alien-looking piglet may actually be a year-old midget.

If you know the breeder, try to observe the pigs a few weeks before you buy so you can choose a pig the way a breeder might choose one for himself. When I want to keep a pig from one of our litters, I not only look for one with the favorable qualities listed above but try to select one that is especially alert, aggressive and competitive for food. As with all animals, take your time when buying. Don't be rushed and insist on choosing your own pig. You can catch a piglet by quietly stalking or cornering it and quickly and decisively grabbing a back leg and reeling it in. Ignore the shrieks and deathlike squeals—they are all just talk.

HOUSING

Pig housing can be as elaborate as you want or your pocketbook will stand, but you can get by quite well with a very small area and minimal expense. Again, start simply. Unless you live in a warm climate, buy a pig in the spring for slaughter in the fall and forget about winter housing.

There are three fencing materials that can be used: wood, wire mesh and electric fencing; and three "confinement systems" (as I call them) for your pigs: small dirt-floored pens, small wooden- or concrete-floored pens and large pasturing pens. Depending on what land you have available and your initiative, some systems offer substantial savings in hog raising; all, however, are suitable for home hog raising. A word first about fencing materials and then more on confinement systems.

Fencing Materials

Pigs are one of the easiest animals to fence, but if they get out, especially if they are young, they can also be the hardest to catch. My preference for ease and economy is an electric fence. We have a small dirt-floored wooden pen for our young weanlings until they are about three months of age when we pasture them in a large electric-fenced field. Pigs can be a problem to fence unless you know what you're doing. If young ones escape without establishing a sense of "where the food is," you may never see them again. Older pigs rarely make an attempt to escape if adequately fed, and even if they do they will not wander far from the site of their daily feeding. We have a friend who lives atop a mountain near us, and after butchering his feeder pigs in the fall he allows his sow to range loose with his sheep.

Electric: This is my favorite because it is cheap, portable and quick to set up. A cheap electric or battery operated fence charger is available for under $20 and a quarter mile of wire and insulators will cost you about $15. Rather than buy metal or cedar posts, either cut your own or go to the local lumber mill and buy (sometimes they are given away) four-foot hardwood stakes for your fencing. A system I use and find quite efficient is ½-inch steel (#4) reinforcing rods with plastic insulators attached by "J" bolts. This is best for portable fencing because the posts can be pulled up quite easily and the insulators can be adjusted in height to compensate for varying terrains and pigs of different ages. The metal rods can be driven into the ground with a hammer or by a simple tool constructed of two 18-inch pieces of pipe and a "T" coupling. The pieces of pipe are screwed into the coupling so they form a straight handle, the hole in the coupling is placed over the rod and you push it into the ground.

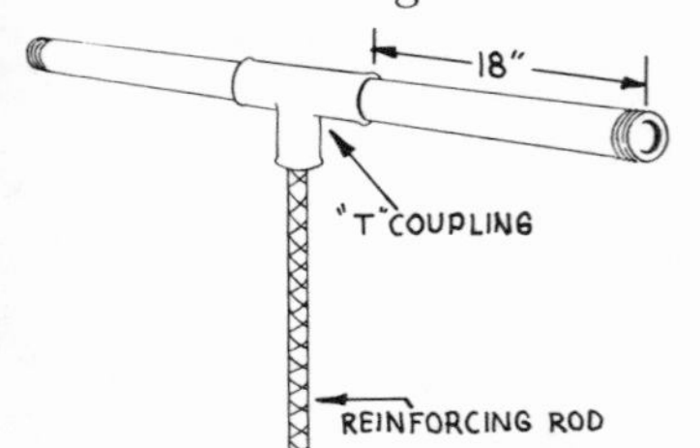

"Pipe tool" used to drive reinforcing rod into ground

Whether you use the metal or wooden posts, they can be pulled up and the whole fencing moved elsewhere to make use of available feed if so desired. Simply set two strands of wire, one at six to eight inches off the ground and the other a foot. One or two inquisitive nudges with a pig's snout and the resulting jolt will turn off the most rambunctious porker. A pig is so smart that after that first run-in it may never hit the fence again, except by accident, which will only serve to reinforce the message. After a year of electric fencing, our sow needs only one wire a foot off the ground, and even then I turn it on only occasionally lest she discover that I have it unplugged. In using electric fencing the charger must have a good ground and you must be vigilant so that weeds or rooted sod do not touch and ground out the fence, thus eliminating the charge. If it does become grounded out, your pigs will soon discover this, and if they are not supplied with enough food will soon make an exit for greener pastures.

Using electric fencing for young pigs may be a problem because the wire is so low that it is easily grounded. For very young pigs the bottom wire should be a bit lower than snout height and the top wire just a bit lower than the pig's height. (Of course, these wire heights must be changed periodically to accommodate your rapidly growing piglets.) We have had little problem fencing even an entire litter of piglets in this fashion. You will find, however, that all pigs are different; some will never go near the wire, while others will go under it all the time unless you use two strands.

Wood: If you use a wooden-fenced pen you will not be able to fence a large area as economically as with electric fencing. It is not portable and it takes more time to put up; however, it is far cheaper than any other form of fencing for a small pen. Don't be fancy—any scrap lumber will do. We had a collapsed chicken coop on our farm and we used the sheathing boards for our first pigpen (cost: nothing). Drive wooden posts into the ground as far apart as your boards are long and build it to a height of four feet. If it's much lower than that, even a 200-pound pig could scale it if so inclined. If you house young pigs, space the boards no wider than their heads (four inches apart or so) because if a pig's head can fit through, the rest of him will have no problem following. Pigs will root holes several feet deep, so you must bury the bottom board a bit to prevent them from rooting out of the pen. Some people suggest ringing a pig's nose or running a strand of barbed wire along the ground to prevent rooting, but I think these are cruel practices. If burying the board doesn't do the trick, try burying some woven wire fencing to the depth of one foot around the perimeter of the pen, or drive stakes along the outside of the pen about 6 to 12 inches apart.

Wire: I think wire mesh, unless found on your farm or inherited, is only for pigs that live on New York's Upper East Side. Currently it is so expensive that if you invest in it you will have the most expensive pork chops this side of the Ritz. If you do have some wire mesh lying around, pen construction is the same as for wood. Make sure the holes are small enough to contain the little fellas and keep it close to the ground or buried six inches and have it at least four feet high.

Confinement Systems

Pasturing: If given good pasture a pig will graze just like a horse or sheep and get up to half its food supply this way. Quite a saving in feed! Using the reinforcement rod posts, we make large portable pasturing pens. Shade must be provided for pigs. Either a large stand of trees or a simple lean-to or A-frame hut should be adequate. If a small stream of water is available to the pig, watering and cooling during hot weather will be simple. If not, you might have to spray your pig or sprinkle it with water during extremely hot days.

We are using our pigs to rototill some rough pasture which we reseed once they have turned it over completely. If you like your pasture the way it is, make larger pens and move the pigs once the grass is eaten down or they will begin to root extensively. To guard against parasites, get into the habit of rotating their pasture, keeping pigs off the pasture for two years after grazing.

Dirt-Floored Pens: If you have a limited amount of land, hogs will do well in a smaller pen, but this will not offer you the feed economy of pasturing. Each pig should have *at least* 100 square feet, but allow 250 or more if possible. While this doesn't afford the extra food that pasturing does, the pigs will still be able to root the ground, augmenting their diet a bit and providing themselves with much-needed iron and other valuable minerals and nutrients from the soil. In a small pen the ground will be rooted bare within a short time, and it is advisable to toss in some fresh sod and/or some fresh grass clippings each day. Again, shade must be provided.

Wooden or Concrete-Floored Pens: By pouring a concrete slab or laying a floor of two-inch boards set a foot off the ground before fencing, you will have a pen that is almost maintenance-free and escape-proof. A quick spray with a hose every few days and your pen will be looking (and smelling) like new. This may seem to be the lazy person's pen—but don't use this system unless you wish to invest a little extra time each day in digging sod and cutting grass. If you simply feed your pig commercial feed and some kitchen scraps each day in a pen like this, I'm afraid you'll get pork not a whole lot different from what you can buy—both in taste and in price. You must provide at least a shovelful of fresh sod *every* day or be content to raise unspectacular pigs.

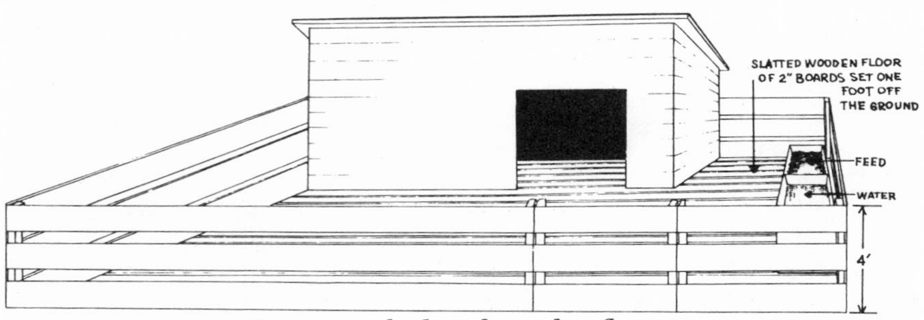

Pig pen with slatted wooden floor

In the case of wooden floors, slat the boards a bit to allow wastes to fall through, but do not space them so far that a pig's foot will also slip through. As with any confinement system, shade must be provided.

I term what is commonly referred to as "confinement housing" winter housing simply because I would never use it except when I have to—in the winter. Commercial hogs are often raised all year round in air-conditioned, humidity-controlled, heated houses with computer-controlled feedings. The method is very efficient and the product will be suitable for anyone who cannot tell the difference between a fresh, vine-ripened tomato and a hothouse one. I don't believe that you can raise a healthy, good-tasting hog, economically, using confinement housing year round. You simply can't discount fresh air, sunshine and pasturing and rooting to get extra food and nutrients. Depending upon your locale, you may not need to confine your pigs even in the winter months. Often a small, draft-free pig house set in your pen will be adequate. We have kept a pig through a Vermont winter with snow and 20 below zero temperatures in such a way by supplying her with extra hay to snuggle up in. I wouldn't suggest this in such extreme climates; we did it because we didn't have time to fix space for her in the barn. She was often stiff and chilled on some below zero mornings, and we may have been flirting with disease.

If you expect to keep a pig indoors during the winter, plan on at least 100 square feet per pig. Allocate an extra 100 square feet for a sow and her litter. The shed or barn should be well ventilated to help expel the large amounts of moisture a hog emits, but it should not be drafty. A low ceiling (preferably with hay stored above) is best for heat-conserving reasons. And the more hogs the warmer! Some publications suggest insulating a hog barn, but I think this is unnecessary (and expensive!). Our house isn't even insulated, and I'm damned if I'll insulate a hog house before ours. As long as your pen is free from drafts and you supply your pig with extra bedding in which to snuggle, it will be fine.

Winter Housing

Tethering: Another form of winter confinement, much like stanchioning a cow, is tethering. It is really only recommended for larger numbers of pigs, but it is worth noting. You can purchase a metal tethering collar, which is secured on the pig and fastened to the ground. The pig is able to lie down and move but is separated from its neighbor by wooden partitions. There is a food/water trough in front and a gutter for wastes behind. I don't particularly like this system because it limits the pigs, but if you have a large herd and are limited in space, I suppose there's little choice.

EQUIPMENT

You will need some sort of feeding dish that will allow at least 1½ to 2 feet of feeding space per pig. It should be nontippable since pigs seem to love to turn their dishes over, wasting food. Old iron sinks are ideal, but while they were given away a few years ago their price now makes them something of a luxury. There is no need to spend extra money on feeding dishes (it ups the price of your meat, remember). Use your head. I found an old coaster wagon and a flat-bottomed barbeque grill at the landfill, and after taking their frames off I had two fine feeding troughs. I do fasten them down because they are so light, but not permanently so that I can remove them for periodic cleanings. A salt block should be available for your pig's free choice.

Pigs need plenty of water. You will not want a huge container because they will dirty the water and it is likely to stagnate. A five-gallon thick, pliable rubber pail or a five-gallon wide-based galvanized tub is best, or any old bucket will do. Try to make use of what you have. The metal ones can be very hard to free from frozen water in the winter, and hard plastic ones will probably not last long, especially in colder weather, because pigs have a tendency, when their pails are empty, to bat them around, carry them into their houses, roll on them and sit on them. After you have exhausted all the old pails on your place buy a flexible rubber one. If you have trouble with your little friends constantly spilling their water, build a wooden frame around the pail or stake it to the ground.

If you pasture your pigs far away from your water source, you might devise some sort of bulk-carrying storage tank that you can fill and transport to their pen, to be stored in the shade and used to refill empty buckets.

FEED

Pigs are like humans—at least in that they have only one stomach; therefore they are not nearly so economical as multistomached animals (goats, sheep, cows) in converting low-grade protein such as forage to meat and other high-grade proteins. Then why are they in this book? For one thing, they are one of the most efficient converters of feed to meat; secondly, they are about the easiest animals for which to secure *supplementary* high-quality food. Because their digestive systems are like those of humans, they need a balanced diet: proteins, carbohydrates, fats, mineral salts, vitamins and water. Table scraps, garden surpluses and good pasture can supply all of these requirements. With a little effort, and some thinking, it is quite possible to raise a pig using almost no commercial feed.

Remember that water is almost as important as feed itself in the raising of a pig. Those pigs that begin to drink water early in life and consume large quantities of it will gain weight with greater feed efficiency.

Commercial Feed

Let us consider commercial feed first because (1) you will probably begin with it, and (2) at times it will be all you have access to. Commercial pig feed is a "complete" feed in that it contains all the food essentials listed above. You can usually buy it in mash or pellet form, pellets often being a bit more expensive. But shop around: local feed stores often have wide differences in prices.

Most pigs like their feed wetted down so it is mushy, but experiment to find out how your pig prefers his food served. Feeding mash without wetting it often makes it unpalatable and may form balls of food in the pig's stomach. In the winter I do not wet the feed, the obvious reason being that it freezes solid and, for some reason, it robs the pig of heat. I find that feeding a pelleted ration, dry, in the winter is the best solution. Keeping the drinking water unfrozen is also important because a pig needs plenty of water to maintain its body temperature.

As mentioned earlier, you will find that two pigs will grow faster (and more economically) than one. Never, when there are two pigs around, will only one be eating. Keep an eye out for "bullying" by one pig. If one pig is getting short-changed because of the other's "hoggishness" feed them at a distance and in separate dishes.

From weaning age to butchering weight (about 200 pounds) you will need 600 to 1,000 pounds of commercial feed or its equivalent. (See *Table 1*, page 163). The most critical time in the raising of a pig is immediately after weaning. If a young pig does not get a good start, it will set a precedent for later growth. Up to 100 pounds a pig needs feed containing at least 16 percent protein; from 100 pounds (about three months of age) to 150 they need 14 percent protein; and from then on to butchering (200 pounds or so) they need 12 percent protein.

Most commercial pig feeds are 16 percent protein so they will be fine for weanlings. At times I have raised the protein content even higher by supplementing their feed with a high-protein feed such as soybean or cottonseed meal. I have done this for two weeks after weaning to give our pigs an extra boost. By mixing the feed equally with corn meal or whole corn after three months (at about 100 pounds) the protein content will be cut to about 14 percent. By increasing the corn content so that at 150 pounds it is two-thirds of the ration you will have a feed that is about 12 percent protein.

A feeder pig (one that you raise for meat) should be fed as much food as it will clean up between feedings. If there is a lot of feed left over from the previous feeding, cut down to avoid waste; if none is left increase until there is just a bit left at the time of the next feeding. I try to feed our pigs three times a day for the first three months and twice a day from then on. This, however, is not critical, and if three feedings are inconvenient, don't worry about it. Avoid feeding young pigs feed that is high in urea as they will urinate out much of the nutrients and make slow gains.

It is not economical to raise a pig beyond 200 to 225 pounds. Greedy ("piggy") people reason that more weight means more pork, but this is not the case. Gains over 200 pounds require more feed per pound of gain than below that weight and gains are more fat than lean meat.

A word about estimating weight: Unless you have a large scale (pigs are reluctant to mount bathroom scales), you'll have to estimate weight. This can be done fairly accurately and easily by (1) using a weight tape or (2) the use of a weight formula. A weight tape, when used to measure a pig's girth, will read not in inches but pounds. You can purchase them in some feed stores or from farm supply catalogs (see *Appendix*). According to Garden Way's Bulletin 18, "How to Raise a Pig," a simple and accurate formula for estimating weight is

$$\text{Weight} = \frac{\text{Girth} \times \text{Girth} \times \text{Length}}{400}$$

All measurements are in inches. The girth is the measurement around the body of the pig just to the rear of the front feet. The length is from the base of the tail to a point between the ears. To make the measurements use a tape measure or a piece of bailing twine later measured against a ruler. If the weight is under 150 pounds, add 7 pounds to the total. An example: A pig measures 40 inches in girth and 43 inches in length.

$$\text{Weight} = \frac{40 \times 40 \times 43}{400}$$

$$= \frac{68{,}800}{400}$$

$$= 172 \text{ pounds}$$

Supplementing Commercial Feed

So much for feeding a pig on someone else's food. While I believe you can raise delicious pork economically on commercial feed alone, a real saving for you (and the world's food supply) can be realized if you supplement or eliminate the use of commercial feed by making use of unused pasture and the tons and tons of perfectly good and highly nutritious garbage that is thrown away to rot every day. It is probably true that what Americans throw away every day could feed half the world's starving people—any pig would attest to that. While my wife worked in a restaurant last year our pig ate better than we did (shrimp, *coq au vin*, veal *cordon bleu*, etc.) on what the patrons discarded.

Pasture: You can save up to 50 percent on feed by putting your pigs out on pasture. Not only is the pasture excellent food, but a pig allowed to "range" can correct any deficiencies in its diet. A pig will eat old trees, hardhack, roots, earth and (beware) sometimes stray chickens. It will extract valuable minerals and nutrients that will enable it to make better use of its food and help keep it healthier and disease-resistant.

You cannot expect a pig to thrive solely on pasture—it is not high enough in protein and does not contain everything it needs for growth. You will still have to supply it with commercial mash, or other foodstuffs, as will be explained later.

A young pig can be put outside as soon as the weather is sufficiently warm and it can be contained by your fencing. In warmer weather a pig can and should be born outside. If already weaned, they also require access to high-quality, high-protein feed. An acre of good pasture will support an average of a dozen young (less than 100 pounds) pigs and six older pigs.

If you have poor pasture, as we did when we first moved to our farm, put your pigs to work. Confine them to a small area of pasture until they've rooted it bare (they will have also gotten out all the weed and quack-grass roots). In extra-rough areas, where there is hardhack and little shrubs, take a stick, and poke a hole around the roots and drop some corn in and—come the next day you'll have a cleared spot. After it's cleared (and fertilized!) move them along and remove the rocks and reseed the old pasture. In time you'll have beautiful, nutritious pasture without having to pay large sums for plowing and reseeding.

Good pasture consists of alfalfa, clover and rape or a mixture. Don't put them on too soon after planting and don't let them eat good pasture down too far or they'll begin rooting extensively—move them on. Good summer grasses consist of Sudan grass, sorghum and soybeans. Good late crops are rye, rye grass and winter barley.

Garden Plowing: If because of hard rains, hail or other catastrophes, you or a neighbor cannot harvest a crop, fence it off and set your pigs out in it. This process will make good use out of an otherwise wasted crop, and save feed costs for your pig. Even after a normal harvest a pig can be fenced in a field to glean wasted crops, stalks and weeds. A varient of this involves the family garden. We have a strand of electric fence wire to keep out varmints and, after harvest, to keep our pigs in. They will clean out any rotted potatoes or crops left behind as well as rooting out weeds and corn stalks. They will till the garden a foot deep and fertilize it to boot.

"Garbage": This is the real harvest. Perhaps I should first explain what I mean by garbage. It is not tin cans, plastic bread wrappers or the rest of last Christmas's mince pie found in the back of your refrigerator in July. It is *fresh* kitchen scraps, plate scrapings and scraps from restaurants or institutions, offal from butchering, stale bread or other foodstuffs that are still edible but not relished by humans, fresh grass clippings, etc. I understand that commercial growers cannot feed "garbage" (raw, at least), but this is our gain. The reason it is prohibited on commercial hog farms is for disease prevention, but in a small operation it should be no problem whatsoever. Disease, particularly and most importantly trichinosis (more about this in the *Health* section), is caused by eating already infected pork or food contaminated with rodent droppings or the droppings from almost any fur-bearing animal (they are all trichinosis carriers). To prevent disease problems in the use of garbage: (1) use garbage before it goes bad and keep it in covered cans or food bins away from rodents and other animals; (2) boil questionable garbage; and (3) feed only what will be cleaned up at one feeding to prevent rodent infestation. Boiling will never be necessary if you don't collect bad stuff and if you exercise a little restraint. That is, if you suddenly get access to half a ton of beautiful garbage, don't take it all. Take as much as you can to fill your containers and that you can use before it rots. The garbage flow will never stop, unless everyone gets with it and into pigs. Our source of "garbage," a restaurant, is unpredictable (depending on their business) and we often freeze surplus in busy periods for use during lean times.

A wonderful program for making use of alternative feeds in raising pigs as outlined by The Small Pig Keeper's Council of England during World War II is as follows:

TABLE 1: Alternative Feed Sources for Pigs *

Weight of Pig in lbs.	*Approximate Age in Weeks*	*Amount to Feed Daily*
30	7	2½ lbs. Grade I foods.
60	12	3½ lbs. Grade I foods.
100	17	2 lbs. Grade I foods plus 3 units of Grade I and II foods in any combination.
140	21	2 lbs. Grade I foods plus 4½ units Grade I, II, or III foods in any combination.

Grade I Foods (1 Unit = 1 lb.)

Commercial pig feed
Kitchen table scraps and swill (providing it does not contain too many potato peelings, vegetables, or too many outside leaves of green vegetables)
Restaurant and institution *"garbage"* (excellent if rich in gravies, meat, piecrust and bread)

* Source: Bulletin No. 18, "How to Raise a Pig," Garden Way Publishing Co.

Kitchen scraps and swill weigh about 2 lbs. to a quart. Solid vegetables and fruits weigh about 1 lb. to a quart. Greens, grasses, vines, etc. weigh about 14 lbs. to a bushel.

Offal [**]
Animal afterbirth (very rich in protein)
Soybeans

[**] Fresh pork trimmings or fat must be boiled first (or frozen for three weeks) to kill any possible trichinae larvae.

Grade II Foods (1 Unit = 4 lbs.)

Potatoes (boiled)
Sugar beets (raw)
Jerusalem artichokes (raw)
Belgian carrots
Corn ears on stalks
Most surplus garden vegetables (except as listed in Grade III below)
Young rape, young grass, clover, fresh clippings

Grade III Foods (1 Unit = 10 to 12 lbs.)
(May need to boil for increased palatability.)

Garden waste, remains of cabbage family, pea and bean vines, vegetable peelings.
Cabbage, kale
Turnips, beets, beet tops
Surplus fruit

As you can see, the higher the food value the less quantity is required. Note the importance of feeding only the best quality, highest protein feed to young pigs. Milk and skim milk are perhaps the best of all pig foods in food value. Milk is also very effective in controlling internal parasites. If you have a family or neighborhood cow or can get surplus milk, it will do wonders for your pigs. Feed up to 1 to 1½ gallons (8 to 12 pounds) per day plus anything above that your pig will consume.

In using supplemental foods you'll have to remember that pigs are similar to humans in digestion. Feed everything in moderation. You would never eat 15 pounds of apples, nor should you feed that much to a pig. Sod fed to a confined pig is an excellent source of minerals as are wood ashes fed alone or mixed with the food. For all pigs you should supply a mineralized salt block, free choice.

The Garden Way bulletin also gives these sources of diet essentials: *Protein:* meat, fish, eggs, wheat, corn, milk; *Carbohydrates:* fruit, vegetables, cereal, milk; *Fat:* meat fats, oils; *Minerals/Salt:* soil, trace mineral salt block, sod.

"Garbage" Sources: These can be as extensive as you want. You might try planting a "pig garden." Soybeans are good feed for pigs and excellent for improving garden fertility too. We threw some squash seeds into a manure pile last year and got pounds and pounds of zucchini for our pig. A word about palatability: if you throw a whole zucchini or other hard vegetable to a pig, it will probably not eat it unless near starvation; however, if you will take the time to puree it, mash or boil it so that it is mushy, your pig will devour it. Also, planting extra corn will pay off, since corn is one of the best feeds for finishing pork.

As far as procuring discarded food, think how many people/institutions use food and proceed from there. Most places will be happy to separate edible from nonedible refuse. As a courtesy supply containers or bags and you will never lose a source. Keep an eye out for community suppers, church suppers and the like for lots of waste. Again, supply containers and offer to help. Schools, restaurants, bakeries, food stores and neighbors are excellent sources. Bakeries often charge for stale bread, but it is much cheaper than commercial feed and nearly as nutritious. If you can be a regular recipient of a restaurant's garbage, you may never buy pig feed again. A teacher friend of ours took a few minutes to draw a sign PIG FOOD and place it over a garbage can he placed in the school cafeteria. Some kids don't like their lunches and throw them away. More joy for your pig—and your wallet.

Sort through all garbage, discarding papers, metal and other inedibles along with bones, coffee and tea grounds, banana peels, and rhubarb tops. Pigs will often discard what is not good for them and what they don't like anyway. You will learn your pig's tastes. Above all, your savings will be in proportion to the time you put into collecting your supplemental food. It need not be a lot—a few minutes on the way home from work can save you many, many dollars in feed costs.

MANAGEMENT

Routines

While buying spring pigs and raising them for fall slaughter is by far the most popular routine, I'd like to say a word about the often unthought of alternative, late summer to winter raising. Granted, in colder weather a pig will only *maintain* weight on what would fatten a pig in warmer months. This means more feed per pound of gain. For that reason this should only be considered if you have a warm pen and a healthy supply of free or inexpensive garbage. On the plus side, piglets are much cheaper in the fall (presently $10-$15 for fall pigs while spring ones fetch $25-$30). A good pig house for use in the coldest months will help keep your pig warmer and cut feed consumption. It cannot be overemphasized that plenty of fresh water is especially important in the colder months for maintenance of body heat.

We find such a system advantageous because we have access to plenty of fine garbage and because we eat two pigs a year but would rather not butcher them at the same time. Many pork products (namely sausage and cured pork products) have a four-to-six-month recommended freezer life, so butchering twice a year allows us plenty of fresh pork.

Handling

If you've ever tried to take your pig for a walk, you'll know why this section is here. A pig has the tendency to go everywhere but where you want it to go, and often great distances in that undesired direction. For young pigs, as in picking your young shoat from the litter, corner it and grab one of the hind legs and corral it in your arms. Holding it upside down by both hind legs often serves to quiet a shrieking piglet.

If an older pig escapes its pen, the problem is minimal because it has a sense of "where the food is" and will not stray far. Younger pigs that have not developed this sense may pose problems. It will help if soon after acquiring your piglet you begin patting and talking to it to cure it of any wildness. This may make butchering that much harder but will enable you to lure back a wayward shoat. If you have a wild piglet and cannot corner it or grab it, do not attempt to run it down—you will never catch it. If all coaxing and enticements fail, begin feeding it out in the yard, gradually moving the dish closer and closer to the pen until one day you will finally have the little tyke eating *inside* its pen; then slam the door and never give it the chance to escape again. Well-made pens and fences will prevent this sort of problem.

Larger pigs that cannot be carried or placed in burlap bags can be moved via a pig box. The box (see Fig. 3) measures 6 feet long, 20 inches wide and 3 feet high and is constructed with a solid floor and slatted sides and top. The door should slide up and down in tracks formed by two ½-inch lengths of angle iron secured on each side of the door opening. There are two handles extending out either end to allow handling by four or more people. Such a box is absolutely invaluable for moving a sow to be bred or taking your pig to slaughter. Building the box with skids as shown will enable you to slide the box and will make movement to a truck or from pasture to pasture considerably easier. A few morsels of tasty food will help lure a reluctant pig into the box, but I have found after a few trips a pig will enter it of its own accord.

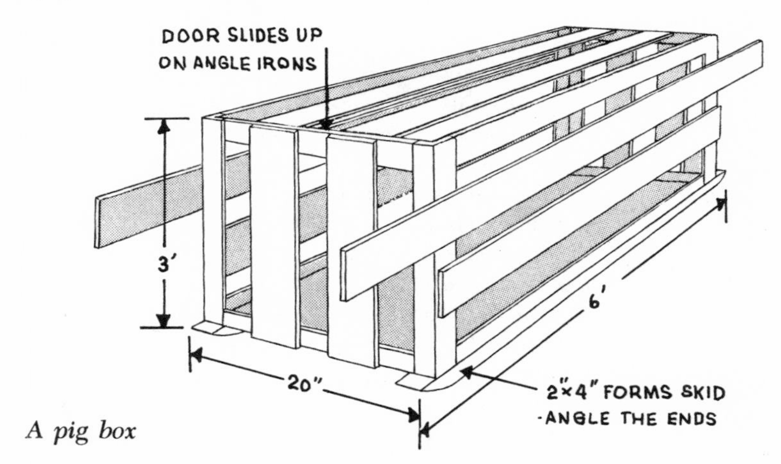

A pig box

An old-time favorite pig-moving technique, and one that I've never had the slightest success with, is jamming a bucket over a pig's head and leading it backward. For some reason when a pig is blinded in such a way it will usually quiet down and permit itself to be led backward by steering the tail. One or more people with sticks can effectively move a pig by "herding." A few cracks of the stick on the side will correct any errors in the pig's navigation. Be prepared for a lot of legwork, since outrunning a pig is not as easy as it looks. Another method of moving a pig, for the stronger and more adventurous, is to tie a rope around a rear leg of the pig and as it runs pull the rear end around so it continues to run in the right direction. The pig supplies the locomotion and you the steering.

BREEDING

While I wouldn't encourage anyone to go into pig breeding the first year, it would certainly be worthwhile considering later. Stretching the mortgage lifter concept a bit, you can keep a sow and after a year or so pay her off by selling piglets, and by the second year begin generating a profit that will enable you to produce free pork for your family. In our area we never have problems getting rid of our piglets and could actually sell many times the number we do. Our first year we bought two gilts, keeping the best for breeding and butchering the other. She had two litters her first year, and by her spring litter the following year she had finished paying for herself (from the day of their purchase until the present) and had begun to show a profit.

A purebred sow is not a necessity but would make your piglets that much more attractive if you can get a good deal on one. Yorkshires and Landraces are prolific and make good mothers, but most any breed or cross will do. A prospective sow should have a medium long body and a strong arched back. She must have strong legs, feet that are well-formed and free from injury and should be alert and active. Look for a large number (at least 12 or 14) teats that are prominent and well spaced.

Unless you plan on a big operation or can expect to make a good income in breeding, you shouldn't keep a boar. It's nothing like sheep, where you can keep a ram for a few extra dollars a year. In our area there are not that many boars, so we have to settle for what's available. A boar should have good conformation and have prominent testicles of equal size. Ask about his breeding record and size of litters. Arrange to breed with the best boar in the area if the price or arrangement suits you (here's an opportunity to barter some of your extra livestock). Contrary to American custom the little lady will probably call on the male. A pig box is indispensable here.

A gilt will reach puberty from five months of age on, but should weigh from 200 to 250 pounds before she is bred. While you can breed a sow at her first heat, larger litters usually result when gilts are bred at their third or fourth heat period. Sows that are in heat may mount other sows and the vulva becomes red and swollen. Often there is a whitish discharge and possibly if you push on her hams from the top she'll look lovey-eyed and stand there waiting. Signs vary with different pigs and they are often hard to detect, but with experience you'll be able to determine when your sow is in heat. The heat period lasts from one to three days (the longer period for older sows) and she'll return in heat every 16 to 24 days until bred.

As with sheep, *flushing*, or increasing the feed intake prior to breeding, will usually increase the number of eggs dropped and hence the size of the litter. While a sow or gilt might normally be receiving four pounds of food per day, during flushing this amount should be increased to five to seven pounds. Access to high-quality pasture is also recommended. You do not want a fat sow, but one that is both trim and gaining weight at the time of breeding.

The gestation period is, conveniently for those of us with poor memories, three months, three weeks and three days (114 days). You can be certain she is pregnant if she doesn't return into heat. Otherwise it may be hard to see any other sign until about a week before, when she'll really begin to balloon and fill with milk. About three weeks before farrowing, when viewed from the side, her teats will appear enlarged and her stomach will sag a bit. When milk can be stripped from the teats, farrowing will usually occur within 24 hours. When a pig is ready to farrow she will become very restless and begin making a nest.

If your sow has good quality, 14 to 16 percent protein feed, no special supplements will be needed. While a sow should be getting about four pounds of good feed per day before breeding and five to seven pounds during flushing, she should be returned to four pounds a day for the first three months of pregnancy. While you want to allow for the growth of the babies, you definitely don't want the sow to become fat. At two weeks to ten days before farrowing increase the feed intake to 1¼ to 2 pounds of feed per 100 pounds of pig. The larger amount should be for gilts who are not yet mature and need feed for growing as well as supporting a litter. On the day of farrowing some people don't feed their pigs, but I prefer to cut her feed in half. Thereafter increase feed one pound per day, giving her as much as she will clean up but not exceeding ten pounds per day. During pregnancy exercise is also important and the sow should be fed as far from her house as possible to encourage movement.

About a week before the sow is to farrow you might want to prepare a farrowing crate (see *Fig. 4*). The purpose of the crate is to confine the babies near the mother immediately after birth and for the first few days afterward without danger of the sow sitting on and crushing them. It is really a pen within a pen. The inner pen houses the sow and prevents her from sitting down too fast or plopping on her side. It is open eight to ten inches from the ground to allow the piglets to go in and

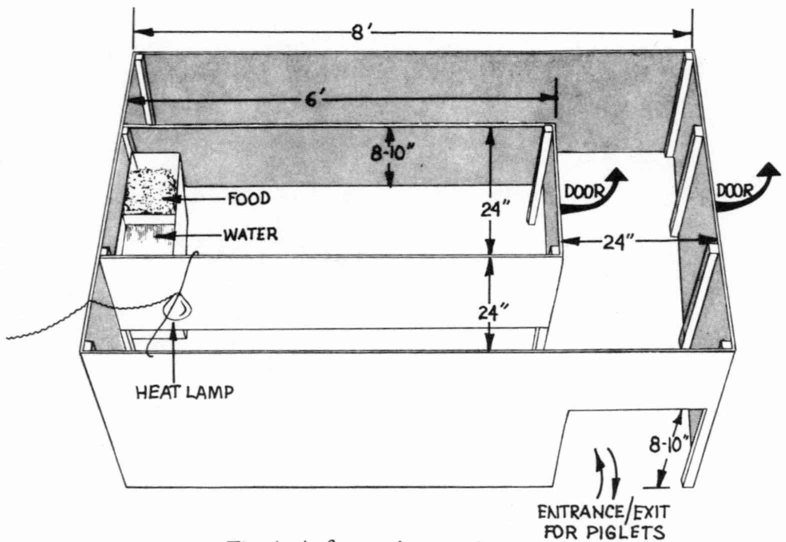

Fig.4. A farrowing crate

out to nurse and to escape as the sow sits down. The other pen is fenced down to the floor and is used to confine the young pigs. On one side of the crate is a heat lamp.

I have mixed feelings about farrowing crates. I think the use of them depends a lot on the sow—and you. If you have a gentle, calm sow and you plan to be around when she farrows, you may not need one. The first year we had a sow, I spent many hours building a perfectly wonderful farrowing crate. Our sow humored us, allowing me to keep her locked in it for a few days before she farrowed, but on the day of farrowing she wanted OUT. And there is no way you'll keep pigs confined if they really want to get out. I let her out with visions of her crushing all her babies, but lo and behold, she did quite well by herself. A lesson perhaps for all stockowners: they did quite all right before we were around and will probably continue to do so if we'll let them. I felt she would do better outside the crate if calm than inside if terribly agitated.

I really think farrowing crates are best suited for larger operations where individual attention and care is simply not possible. If you do not have a farrowing crate, you should have a pig brooder four feet square with a heat lamp set in a draft-free corner (and install guard rails on the remaining sides of the pen. They should extend eight to ten inches from the wall and be eight to ten inches from the floor. They permit the piglets to move about without danger of being crushed between the sow and the wall.

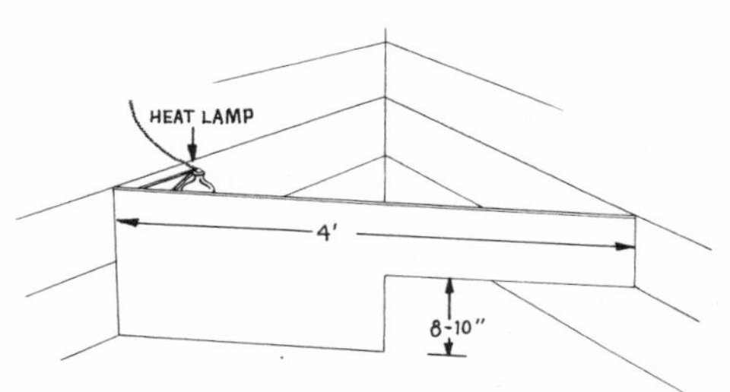

A pig brooder

Once the sow appears to be near farrowing she should be locked in the farrowing crate if you plan to use one. Especially if it is her first litter, you should be around. But do not make too much noise and certainly do not bring in the entire neighborhood to witness the event. A sow can become nervous with a lot of commotion and may eat her young or even hold back farrowing until she is alone. If you have tamed your sow well, it will pay off many times over because she will probably allow you to be present and help in case of problems.

Once labor begins it should take about an hour or two, but exceptions abound. (Our sow has taken as long as six hours.) If she has one or two piglets and none for an hour or so, or if she is in labor for a long time without having any at all, leave her alone for a while. Often sows get bashful and actually hold back while people are present. If this still doesn't produce results within an hour, call a vet and explain all details. The vet may prescribe an injection of a uterine contracter to aid delivery, or may decide that she will need a hand (literally) in farrowing. At times a piglet will become lodged in the birth canal and a little help will be needed. Wash well before and after (before to protect the sow, and after to protect yourself) and coat your hand with Vaseline. Bend your fingers at the knuckles so that your fingernails don't stab her and insert your hand (you'll go pretty deep). Feel around and don't be surprised at the kicking or even nipping. You should be able to turn the infant around and gently pull it out. Either direction will do as pigs are born both head and feet first.

I like to be present in colder weather to dry off the babies, let them get some colostrum and place them under the heat lamp. A heat lamp should be used when the temperature drops below 50 degrees. In warmer weather our pigs are born outside and I generally leave them alone. You should remove the afterbirth, because it is not necessary for her to eat it and doing so might even encourage her to devour her young. It is very rich in protein and can be fed to other pigs (not other sows). If the navel cord is long and unduly hampers movement, clip it to three inches and swab or spray with iodine. I usually don't bother with it; it will shrivel and fall off in a day or two.

A mother will usually accept her litter and the babies will milk about every two hours. If she refuses to allow them to nurse, give her a little tap to encourage her to cooperate. If this doesn't work and you are sure she won't have them, try giving her a pound of salt pork or some beer and then try again—she'll be too sick to care. If she still won't have them you'll probably lose the litter, unless you want to stay awake a few nights feeding them. If you do, goat's milk is a good substitute for sow's milk. A bottle with a nipple will not be needed because piglets can be taught to drink from a pan soon after birth. Feed them every two hours for the first three days, six times a day for the next four days, and thereafter try to keep fresh milk in front of them at all times. Allow plenty of space at the feeding bowls and be sure that they are heavy or the pigs will spill the milk continually. In a day or two after birth you can offer them some grain moistened with milk; allow this free choice. Water should also be available at all times. If the piglets did not get any colostrum, try to get some from the sow and feed a tablespoon to each piglet through an eyedropper. If no colostrum is available, add some cod-liver oil and mineral oil to their first feeding for its vitamin and laxative effects. If you happen to have another nursing sow, you might try foisting a few on her. (You can try this both in the case of an orphaned or rejected litter and with oversized litters in an effort to "even off" the litters.) In the case of a nervous sow who does accept her litter, enter her pen first the first couple of days only for feeding and do not organize tours with the neighborhood kids.

If you use a farrowing crate, you can let the sow and litter out of it in three days to a week, still, of course, allowing them access to the heat lamp. The light performs the double purpose of keeping the piglets warm and attracting them away from the mother after milking and lessens the chance of any being crushed. As the weather gets warmer and/or the piglets get older, the light should be extinguished periodically to gradually wean them from it. Even if the light is not turned off, I have found they will usually wean themselves and begin sleeping outside "where the big pigs do." Young pigs are really quite amazing, being able to scuffle and run around an hour after birth. The first day or two are the most critical in terms of being crushed, and guard rails and the removal of heavy bedding or nest material (which the piglets might bury themselves in, out of sight of the sow, and be inadvertently crushed) will go a long way to preventing losses.

Within three or four days you can allow the piglets access to some pig feed mixed with warm skim milk. This is especially important with large litters to take the edge off their appetites and give the sow a break. If any runts in the litter are having difficulty getting milk, we often supplement their feed with one to three feedings per day of warm skim or goat's milk mixed with a little grain. We usually slip it out of the pig house and run it into our kitchen and give it the royal treatment. We have saved many a little pig this way that I'm sure would never have made it without the little extra strength we gave it. Water given at an early age has been found to be critical for later growth. Pigs that are not given access to water early in life will learn to do without it but may require up to 25 percent more feed for growth! Supply plenty of clean, fresh water.

If you notice that the piglets are cutting the mother's teats or are nipping each other in competing for food, it may be necessary to clip their needle teeth. I don't believe in doing so unless it is absolutely necessary since it may hamper their early eating of solid food, and because clipping may injure the mouth and gums, inviting disease. It is usually only necessary in larger litters where competition for milk is fierce. There are four of these sharp teeth on each jaw, and if they must be clipped use a sharp toenail clipper or cuticle scissors and neatly clip only the tip off.

Males should be castrated at five days of age, because the stress is less and the mother's milk is then at its peak. There is rarely more than a trickle of blood and recovery is quick. While I think it's best to watch the operation before trying it yourself, it is quite easy: have on hand a new razor blade or surgical knife and a spray can of merthiolate or other antiseptic. Wash your hands well. Have someone hold the pig upside down by the rear legs with the pig's head between his knees. After a few minutes of being in this position the pig will quiet down. Make an incision between the two testicles or above each one until they are exposed. Cut through the thin membrane that contains them and push the testicles with your finger until they fall forward. Sever them from their cord and spray the wound with antiseptic. Check the wound every day for a week to guard against infection.

Iron is essential in preventing anemia in young pigs. A sow's milk does not contain enough and even if the piglets have access to sod they will still not absorb all the iron they need. Inject 1 to 1½ cc. of iron in the fleshy part of the pig's neck at two to three days of age. Injecting in the neck is preferred over the rear leg to prevent lameness and staining the hams. Iron shots will reduce mortality, boost weaning weight and enable the pigs to gain faster on less food.

At four to six weeks wean the piglets from the sow. Wean the largest first and leave the smaller ones on longer to give them an added boost. You may have problems with the larger piglets accepting the late weaners, so be prepared to house them separately or else wean them all at once. Try to wean them so that they are out of sight and earshot of the mother and vice versa. This is mostly for the piglets' sake because after four or five weeks of them the sow will be glad to get rid of them. Cut the mother's food down to normal maintenance ration at this time and supply her with plenty of roughage to aid in drying her up. In about 21 days she will come into heat and can be rebred if so desired.

HEALTH

Most hog diseases are reserved for large-scale hog-raising operations. That is to say, a few pigs that are not overcrowded and have access to adequate food and sunlight should give you few problems. Ours have never had more than a few colds and lice, but it is helpful to be able to spot the diseases early so prompt treatment can be given. A good veterinarian book (see *Appendix*) is invaluable; and there is also an outline of hog diseases in the *Appendix*.

Here is the McLean County system of swine sanitation:

(1) Clean farrowing pens thoroughly and scrub them with scalding water and lye.
(2) Clean the sows, particularly the udders, just before farrowing.
(3) After farrowing haul the sow and her litter to clean pasture.
(4) Keep the pigs on this land until they are four months old.
(5) Practice pasture rotation. Keeping pigs off pasture for two successive years.

Be watchful of young piglets who do not compete well for the sow's milk or seem off their feed and cough and shiver. This may indicate pneumonia or strep throat and should be treated with 1 cc. of intramuscular penicillin (in the ham or neck) twice a day for five days.

BUTCHERING

By the time a pig is five to six months old and about 200 to 225 pounds it is time to have it slaughtered. All good friendships must come to an end. Although paying someone to butcher your pig will up the price of your pork, it is advisable to have it done at least once and *watch* so you can do it in the future.

Slaughtering a pig is no small operation to be taken care of between halves of a football game. The pig must be shot, stuck (bled), dipped in hot water and have its hair scraped off. Then it is gutted, split and hung. After the meat is thoroughly chilled the carcass is cut up into chops, ham, etc. There are several good books dealing with all aspects from slaughter and butchering to curing meat and smoking, and making sausage and lard (see *Appendix*). Whoever does it, however, and whatever your ultimate cost, it will be the best pork you've ever tasted.

CHAPTER SIX

Veal

If you are tired of paying the astronomical prices for veal (or have you simply not tasted it in years?), you might be surprised to learn that you can grow 100 pounds of veal in just three months. No other animal we are talking about, including pigs, can give you that much meat in a single-animal operation in such a short space of time. While you can do it economically on commercial feed alone, it will be substantially cheaper if you are able to raise some grain or especially if you have access to good cow's milk.

Veal calves require an absolute minimum of space and time and could conceivably be raised in a surburban garage without anyone knowing, if you could explain the daily supply of rich manure you have for your garden and the occasional unusual sound emanating from within (car trouble?). Alas, there are problems. This furry little ball must be butchered in three months. Remember that by keeping his pen clean and feeding him well you are giving him a nicer life than he would have elsewhere. Above all, do not make friends with him. A veal calf you decide to keep will soon eat you out of house and home (and break out of your garage if you are keeping him there). Be comforted also by the fact that such a good source of manure can taste so good.

Holstein calf

BREEDS

While any breed will produce good veal, Holstein is preferred. A cross of a Holstein and a meat breed is also a good choice.

PURCHASE

In purchasing a calf the biggest consideration is not necessarily *what* you buy but rather *where* you buy it. Livestock auctions are among the worst places to purchase calves. The majority of the calves are a day old. Being taken away from their mothers, transported and crowded is often too much of a shock. A 50 percent mortality rate with such day-old animals is not uncommon. While they are more expensive, older (preferably week-old) calves are worth the extra money. If they have made it to that age, their chances of survival are considerably better. However, you still don't know how the animal was fed or handled from birth. Your best course is to make arrangements with a local farmer to buy one of his bull calves (heifers are fine for veal but dairy farmers usually keep heifers for replacement stock or sell them at a higher price). They often have trouble getting rid of bull calves and you can usually get one cheaply. Make sure the calf gets colostrum from the mother for three days after birth, and see if it can possibly be fed the milk for a week before you pick it up. In most areas farmers cannot market milk from cows until a week after they have calved. This milk would then go to waste but it can be fed to a veal calf. These animals will be healthier and more thrifty to feed.

Wherever you buy your calf, select one with a long, blocky body. Avoid those that are narrow and shallow-bodied and that have long, crooked legs. You want alert calves that appear healthy and vigorous. Watch the calf for a while, try to see some of its manure. If it is pasty and white it probably has scours and should be avoided. Check the navel to be sure there is no infection there. Calves need not be castrated or dehorned because they will be butchered before these features matter.

HOUSING

Housing for a veal calf should be no problem. The chief requirement is that the calf be kept dry and free from drafts. You can probably make use of some space you already have without having to build a pen. You can use a lambing pen, a horse's tie or box stall, a corner of a garage or barn, etc. Our pigpen can be used in the late spring or early summer after our pigs have been put out to pasture. A veal calf should not walk around a great deal, and while tying is not necessary,

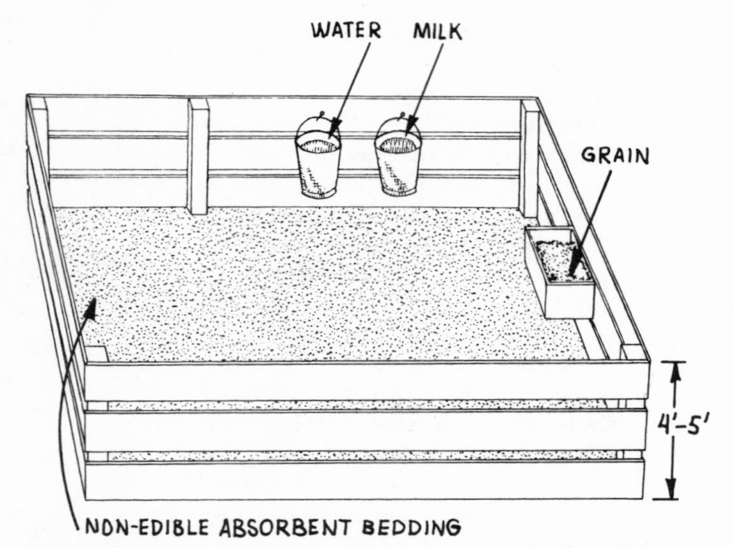

Veal calf pen

you should limit the size of the pen. Ventilation is important in the summer so the pen does not get too hot. If you have windows, they can be taken out and the space covered with burlap. This will allow air to pass through but will keep the stall dark enough to discourage flies. Do not allow your calf to eat anything but milk or milk replacer and a little grain; make sure no pasture or hay is available. The pen should be well bedded, but do not use any edible bedding such as old hay, straw, ground corn cobs or corn stalks. Use wood chips, sawdust or any other nonedible bedding. A pen must be thoroughly cleaned before a new calf is placed in it.

EQUIPMENT

Equipment is also minimal. You will need a dish for feed, one for water, a salt block offered free choice and a container for feeding milk. You can use a bottle with a nipple, but in time that will not hold enough milk for a feeding, so the best bet is a "nipple pail" (sold commercially under many names, among which "Calf-a-teria" is the most fitting). This will hold a large supply of milk. After each feeding, the milk pail and nipple should be thoroughly cleaned and sterilized with hot soap and water. Hang the bucket or bottle so the calf must raise its head—the more natural way to drink. Nipple feeding eliminates waste and also cuts down on gulping, which can lead to digestive disturbances and scours.

FEED

Veal is often raised from birth to butchering on milk (or milk replacer) alone. Such an animal can be anemic, terribly fragile and disease-prone. We believe that feeding milk in the recommended quantities and allowing a grain mixture free choice will produce a healthier, hardier animal without being detrimental to the meat itself. The meat is still pink in color and fries up white and tender.

So-called milk replacer, which when mixed with water attains the consistency of liquid milk, is sold under many brand names in feed stores. Some companies market a milk replacer made especially for a veal calf, but you can raise your calf on the common milk replacer just as well. Follow the manufacturer's directions for mixing the powder. Usually it is mixed with a wire whip with 115° water and served at a temperature of 105°. When the calf is four weeks old consumption is greater when the milk is served at 85°. Feed twice a day as close to 12 hours apart as possible. If the package directions do not give adequate information for feeding veal calves (as opposed to nonveal calves that are being fed grain and pasture or hay), supply them with as much milk as they can finish off in 10 to 15 minutes. Do not overfeed, especially in the first few days of life (see *Health* section). The feeding of fresh milk will be covered in the section on *Supplementing Commercial Feed.*

A commercial or home-mixed grain comparable to 14 to 16 percent dairy fitting ration should be available free choice at all times. With their milk ration the calves' grain consumption will not be very high but will be enough to make them a little hardier. Water and a salt block should be available at all times.

If the calf eats slowly or is off its feed, check his temperature and his manure. While the manure will be looser than that of cows or calves on conventional feed, if abnormal runniness is present follow this schedule:

TABLE 1: Feeding Schedule For Calves With Runny Manure

Next two feedings: withhold all feed.
3rd and 4th feedings: normal amount of liquid, but half of normal solids. No grain.
5th feeding: normal schedule.

A teaspoon of sugar in each feeding tends to reduce diarrhea.

Supplementing Commercial Feed

A grain mixture of your own grains will naturally save you some money, but the real savings can be made if you can "replace the replacer" with real milk. This can come from your own cow if you have one or from local cow owners or dairy farmers. Dairy farmers cannot usually market milk from a cow until a week after calving. If you can make an arrangement with a local farmer to buy his unsalable or surplus milk, you can make substantial savings. Fresh milk is naturally much better for calves and scouring is much less common when it is used. Feed at the temperatures recommended in the previous section and feed as much as it will clean up in 10 to 15 minutes twice a day. Stick to milk from the same farmer, since switching back and forth between flocks can upset the calf's digestive system. Also, try to stick to either milk replacer *or* fresh milk; do not change feeds often but if you must (should the supply of fresh milk run out and you have to switch to replacer), do so *gradually* by mixing a little of the new feed into the old until it gradually replaces it.

MANAGEMENT

Routines

You can set your veal-raising schedule to the time when a local farmer's dairy herd is freshening to make use of unsalable milk. This may also be a time when he can furnish you with a bull calf. It is possible that a farmer may give you two calves and enough surplus milk to raise them in return for your raising one, butchering it and wrapping it for him. If enough milk is available, or even if you have to use milk replacer, you can raise two calves and the sale of one will pay for the other (and maybe then some). If you do raise more than one calf, raise them in separate pens.

The calves tend to grow better at temperatures between 50 and 60 degrees so you can also set your raising schedule to the times when those temperatures are likely to occur in your region. You might also be able to realize some monetary gain in the sale of hides to a local tanner or by selling the hide that you tan yourself (see *Appendix*). With the hair on, calf skins make fine rugs or bed coverings; without the hair their leather is of high quality.

Handling

Handling of veal calves is not a great problem because they rarely get larger than 200 to 250 pounds. A halter or collar will make handling easier and they can be made to follow by coaxing with some grain. A sharp crack on the butt with a stick will also get them moving in the desired direction.

HEALTH

The major disease or parasite problems are outlined for easier diagnosis and treatment in the *Appendix*. Again, strict sanitation, correct housing and regular feeding will go a long way towards preventing disease. The most common affliction, scours or infectious diarrhea, is treated in more detail in the following paragraphs.

Scours is responsible for 10 to 15 percent of all calf deaths. Not to be confused with ordinary diarrhea, calf scours is caused by a bacteria and after symptoms appear death usually results in 24 to 36 hours. The bacteria is present in healthy animals but flares up in calves perhaps after another virus breaks down body resistance. The symptoms are a whitish-yellow, foul-smelling diarrhea. It usually occurs in calves less than three weeks old; they become dull and listless, lose their appetite and their temperature goes below normal. The most common source of infection is through the mouth from sucking a contaminated udder or contaminated objects.

Scours is associated with a lack of vitamin A. The colostrum, which is rich in vitamin A, is an important source, and this is why day-old calves that haven't had enough are prone to scours. Again, three days of colostrum feeding is essential. Strict sanitation is also important in prevention. Also, overeating, which may cause indigestion and lower resistance to disease, should be avoided. Feed only the recommended amounts, especially early in life.

In the case of the infected calves, isolate from any other cows or calves, and clean the pen thoroughly. Call a veterinarian; he might prescribe antibiotics as they have been effective in treatment.

BUTCHERING

When the time comes, you will find that not having made friends with the little fella will make the task much less of an ordeal. The optimal butchering age is 9 to 12 weeks. If your food supply will allow, you might try a few more weeks of feeding, but beware—if you allow the calf to get too old the meat will lose the quality of veal. At 8 weeks the average weight for a Holstein calf is 160 pounds; at 12 weeks, 215 pounds. At these respective ages Jerseys weigh 102 and 141 pounds and Guernseys weigh 122 and 170 pounds. You can see why Holsteins are preferred. The calf should dress out to about 40 percent of live weight for the Holstein (slightly more for other breeds since the Holsteins are bigger-boned), or about 90 pounds of veal for a 12-week-old calf.

We cannot fully describe slaughtering and butchering here. A calf, except for boning of the legs if desired, is not a difficult animal to butcher, and watching the entire process carried out by an experienced person is the best way to learn.

That little corner of your shed that once housed your lawnmower (before you traded it in for sheep) can be put to good use in the production of your veal. Like lawnmowers, four-dollar-a-pound veal can also be a thing of the past.

Don't forget sweetbreads!

CHAPTER SEVEN

Grow Your Own . . .

Throughout this book I have mentioned mixing your own feed rations, growing your own hay, seeding your own pastures, and in general using whatever you have the space and time to raise. To leave it at that, I think, would be negligent on my part. While growing your own feed is the subject for another book, I hope to touch here on some major points and, most important, offer a good body of references you can turn to (*Appendix G*). Furthermore, any large-scale farm which grows its own hay, corn and grain requires expensive machinery (which is beyond our scope), but a lot of what follows can be done with hard work, sweat, and time-honored methods. You won't find a Kansas combine on a Vermont hill farm, but you usually do find a neighborly farmer who will bring in his equipment in return for your help with his haying.

What we are concerned with are *field crops* and *forage crops*. Field crops are those plants that are primarily grown for their seeds: corn, wheat, oats, soybeans and even sunflower seeds. Forage crops are the plants or parts of plants that are used for feed before maturing or developing seeds. These forages are fed as pasture (the easiest way—the animal does the harvesting itself and there are no curing or storage problems), as hay or as silage. The forages are broken down into legumes and grasses.

FIELD CROPS

Corn

Of the five types of corn, flint and dent corn are those that should be considered for feed purposes. Dent corn is the most widely grown feed corn but flint corn, because it grows more quickly, is used in regions with shorter growing seasons.

Plant the corn in rows 40 inches apart, with the seeds planted 5 to a hill and the hills 40 inches apart. Do not pick feed corn as you would your garden variety sweet corn. Wait until after a frost when the husks are dry and the kernels firm. For small-scale operations, hand picking is the rule. A husking tool will make the job faster. The stalks may be saved as they make good bedding for other stock. We feed the whole ear (sometimes the entire plant) to the pigs. For other animals you must husk the corn and then remove the kernels from the ear with a corn sheller. (A hand sheller can be purchased inexpensively from farm supply catalogs or from local feed stores.) Whole corn can be fed to stock as indicated or it can be cracked or made into mash and mixed with a particular feed ration. Corn, plant and all, can also be used for silage. *Yield:* 75 to 100 bushels per acre.

Oats Plant your oats two inches deep and four inches apart in rows that are also four inches apart. Oats should be harvested in the so-called dough stage when they are soft but not mushy. The grain heads should be full but not so dry that they'll fall out. Cut with a scythe, preferably with a cradle attachment to catch and help pile the oats, and rake into windrows and let dry for a day or two. Then tie them in bundles and let them cure in the field for as long as it takes for them to be thoroughly dry (two days to two weeks, depending upon the weather). Then the grain must be threshed (the grain separated from the plant) and winnowed (the chaff separated from the grain). To thresh, place the grain 6 to 12 inches deep on a clean floor (or on a tarp on a not-so-clean floor) and beat the plants with a flail until the grain is separated from the plants. (A flail is a wooden rod about five feet long with a joint made of chain about ⅔ of the way down.) After flailing, take a pitchfork and scoop out the straw (good for bedding), being sure to shake it well and release any oats left in it. Winnowing is not necessary for oats that are to be fed to rabbits as they will husk the oats themselves, but must be carried out for other stock. To winnow, take the oats outside on a windy day and shake them from a bucket onto a tarp while standing on a stepladder. The chaff will blow away and the heavier oats will fall onto the tarp. Be sure it is not too windy or you will be sowing your domestic oats wastefully. *Yield:* 30 to 50 bushels per acre.

Soybeans Soybeans are very high in protein but are somewhat less palatable than other stock feeds. They are grown very much like garden beans but require a longer growing season. Plant six inches apart in rows that are three feet apart. Allow the bean to mature fully so it is hard. Soybeans can be fed whole if the animal will eat them or ground as a component of a feed mix.

Sunflower Seeds

Sunflower seeds are a high-protein seed that are quite easy to grow and make excellent food for chickens. Simply sow the seeds where you want these tall plants to grow, allowing plenty of room between each plant. Let the flowers mature and "go to seed." When the seeds are dry, remove them from the flower head by rubbing it face down on some ½-inch hardware cloth. The seeds can be fed whole as chickens will shell them themselves.

Wheat

Get a variety of wheat earmarked for livestock feed purposes. Wheat can be sown by hand, as depicted in the famous painting by Millet, *The Sower*, or by use of a seed broadcaster, which can be purchased cheaply and will spread seed evenly as you turn the crank and walk down the plot.

Harvest is similar to oats, and threshing and winnowing are also necessary.

Storage

Grains can be stored in barrels or other covered containers free from both rodents and moisture. Make sure the seeds are dry enough (10 to 15 percent moisture content) before storing, or they may get moldy.

FORAGE CROPS

Pasture

Good pasture will enable you to fatten and maintain your livestock with a minimum of grain. The chart below lists common forage crops and the regions most suited to their growth.

Don't overgraze your pastures. If you have animals that will graze very close, such as sheep and goats, have other pastures so that you can rotate them and let the other regrow. Don't put your stock to pasture too early in the spring, or it will never get a good start. Check with a local feed store, nursery, or your extension agent to determine what pasture crops are most suited to your locale and particular livestock needs. Good pastures (and hays) consist of a legume/grass mixture. The legumes supply protein while the grasses are high in

Most Abundant Forage Grasses and Legumes by Region

North Central & Northeast

Grasses	*Legumes*
Kentucky bluegrass	white clover
timothy	Korean lespedeza
redtop	sweetclovers
orchard	alfalfa
Canada bluegrass	common lespedeza
tall oatgrass	red clover
meadow fescue	alsike clover
smooth brome	hop clover
sudan	black medic
ryegrass	crimson clover
bents	birdsfoot trefoil
beardgrasses & bluestems	

Southern

Grasses	*Legumes*
Bermuda	common lespedeza
carpet	hop clover
dallis	white clover
vasey	Persian clover
redtop	black medic
Bahia	spotted bur-clover
fescue	
Rhodes	

Great Plains

Grasses	*Legumes*
grama	sweetclover
wheatgrass	alfalfa
buffalo	
brome	
galleta, tobosa, curly mesquite	
bluegrass, bluestem	
sudan	
bluegrass	
timothy	

redtop
fescue
Rhodes

INTERMOUNTAIN

Grasses	*Legumes*
wheatgrass	alfalfa
grama	white clover
brome	sweetclover
galleta, tobosa, curly mesquite	alsike clover
	red clover
	black medic

SOUTH PACIFIC COAST

Grasses	*Legumes*
fescue	California bur-clover
brome	alfalfa
wild oats	white clover
Bermuda	black medic
sudan	

bluegrass	California bur-clover
timothy	strawberry clover
redtop	
Bermuda	
beardgrass or bluestem	

NORTH PACIFIC COAST

Grasses	*Legumes*
ryegrass	red clover
Kentucky bluegrass	white clover
bent	hop clover
reed canary	alsike clover
orchard	
meadow fescue	
redtop	
timothy	
tall oatgrass	
meadow foxtail	

(From USDA *Yearbook of Agriculture*, 1939.)

energy. Also, the grasses will hold the soil better and prevent erosion and fill in well between the less dense legumes.

Comfrey is unexcelled as a low-cost, fast-growing perennial that makes a highly nutritious livestock feed. Kale and collards also make good feed. These plants can be fed to poultry, sheep, rabbits, goats, pigs—most any livestock. They are very high in vitamins and minerals and can even be fed to young animals as they do not cause bloat the way many other succulent green forages may.

Hay

Hay is simply your pasture cut, dried to about a 15 percent moisture content and stored, either baled or loose, for winter feeding (or whenever pasture is scarce or not available). A legume/grass hay mixture is best, for the reason outlined above. But this is definitely not to say that you must plow up your fields and reseed them. Recent studies have shown that a well-fertilized field of grass, properly cut and cured, will have nearly the same nutritional value as legume/grass hay. This again may make it cheaper in the long run, because legumes generally have to be reseeded more often than grasses.

While all the talk in agriculture circles on hay-making is about balers, curing hay in the barn with hot air, and even balers that make three-ton bales, the only economically feasible method of producing hay on a small scale is by hand cutting, curing in the sun, and piling the hay loose in your barn. Second-cutting hays (because they are more tender and have less hard stems) make the best feed and less is wasted. Most of the nutrients are concentrated in the leafy portions of the plants so hay that is cut when the plants are most leafy and are from well-fertilized fields offer the most nutrition. The table below illustrates the importance of an early cut and proper curing. It can be used as a guide both when making your own hay or when choosing hay that is offered for sale.

TABLE 2: How To Determine Hay Quality—Cornell Study

(1) *Stage of growth when cut* (the more mature, the more loss of nutritive value)

	Digestible Protein	*Energy as T.D.N.*
a. vegetative stage	18.7%	70%
b. bud stage	14.5%	63%
c. bloom stage	10.2%	56%
d. mature stage	6.4%	49%

(2) *Leafiness*—important (more proteins and vitamins in leaves)

(3) *Green color*—very important

a. need 35%-60% of original color
b. lose color—lose 90% vitamin A
c. one or two rains on dry hay causes loss of 40%-60% feed value

(4) Compare alfalfa, a legume hay, and timothy, a grass hay, as to feeding value. Notice that the later grasses are cut, the lower the protein content.

	Digestible Protein	*Energy as T.D.N.*
a. early bloom timothy	4.2%	51.6%
b. full bloom timothy	3.2%	48.0%
c. late bloom timothy	2.4%	47.5%
a. leafy alfalfa	12.1%	51.1%
b. good alfalfa	10.3%	51.1%
c. stemmy alfalfa	8.2%	47.5%

(From: *Dairy Goats, Breeding/Feeding/Management.* New England Dairy Goat Industry Leaflet 439.

As recommended above, cut your hay when it is in the very leafy, vegetative stage. Use a good scythe and have a sharpening stone handy to keep the blade razor sharp. Choose a good warm, sunny dry day to do your cutting and try to ascertain as best you can beforehand that the next day or so will not be rainy. Your speed of cutting with the scythe will, of course, increase with practice and most important depends on a sharp blade.

After cutting your hay, pile it in windrows as with oats and wheat. Curing time depends on how hot and dry the weather is. This past summer hay cut in the morning was cured and ready for baling by midafternoon. But in some cases it will take two days or more. Heavy dews or drenching rains will rob your crop of nutrients, so try to get it in the barn before wet weather. (Bank on some long, hard days when you make your own hay. If you've cut and dried your hay by nightfall, you probably won't get the long night's sleep you think you'll need before loading it and taking it to the barn. Often if the weather report is bad, you'll work late into the night to get your precious crop in before the first drops of nutrient-robbing rain fall.) Turn the hay in your windrows when the top hay dries to get the bottom cured. The hay should not be so dry as to be brittle, but you must be very careful it is not too wet when you store it or it will get moldy or, worse, it will start a fire by spontaneous combustion. Do not overcure, as dry hay is unpalatable and has lost many of its vitamins and nutrients. Check with a local farmer and have him show you the correct condition for your hay before you put it in the barn.

Stack the hay on a wagon or in the bed of a pickup truck (with practice you'll be amazed at how high you'll be able to stack the hay and have it stay together) and then unload it in your barn. Pile it on a wooden floor or on a dirt floor covered with old boards. This is to prevent the hay from getting wet and rotting.

Silage

Silage is such crops as grasses, legumes, corn (plants as well as corn) and the like, chopped, stored and fermented in the absence of air. After the silage is packed into the silo or other airtight container, the oxygen is used up in a short time and anaerobic bacteria thrive which in turn produce lactic and other acids. As the acids increase, the bacteria die off and, as long as no more air is permitted to enter, the silage will keep almost indefinitely.

While the large silos attached to barns are most common for making silage, these are much too large for the small-scale operation we are talking about. Since the most important consideration in making silage is the absence of air, any reasonably airtight container can be used. Fifty-gallon drums are a good size and the most available. Since silage will begin to spoil as you use it, the use of a number of drums will enable you to use one container at a time and not risk spoiling all your crop. Of course, any other airtight container, smaller or larger, can be used to suit your needs. If you really get into making silage and find it affords you good feed savings, you might want to construct a larger, more permanent silo. A large circular silo (square containers are less often used because of the difficulty of eliminating all trapped air from right-angled corners) made out of corrugated steel roofing fitted with a tight top and set on a concrete slab can be made easily and cheaply. If you place an airtight door at the bottom, you can remove the silage with a shovel without worry of spoilage.

Whether you use drums or other silos, the procedure for making silage is the same. The moisture content of the material to be ensiled is very critical. It should be between 55 and 65 percent moisture (as a guide, freshly cut hay is about 75 percent moisture); too moist and it will spoil easily, too dry and it will be unpalatable to your livestock. In order to pack and ferment well, the material should be chopped. Farmers use a silage chopper, but you will have to do this by hand or use a shredder or composter if you have one. It is best to have it cut into ¼-inch lengths or be finely chopped. Pack it tightly into your silo and fit it with a tight cover. Every day for a week, stomp on it and press down to exclude any trapped air. Do this until the mixture is settled. Be sure the top is on tight and you should have a minimum of spoilage. Some people like to add molasses as a preservative but this is not essential. Your silage should be ready in about three weeks.

You will always have a little spoiling where air has leaked in after you begin removing the silage. Be sure to remove any moldy or spoiled silage before feeding it and check your batch each time you remove some to be sure there isn't any undue spoilage.

Seeds And Fertilizing

Buy your seeds locally; these will be most adapted to your particular growing region. Check with local feed stores and, more important, your local extension agent, to determine what crops and varieties are best suited for your region.

Have your soil tested before sowing your crops and periodically every few years after to determine what is needed in your soil. Most state universities will test soil samples free of charge or for a nominal fee. They will recommend the correct amounts of chemical or organic fertilizers, depending on which you state as your preference.

Appendix

Appendix A.

Animal	Female	Male	Young Animal	Castrated Male	Normal Temp. °F.
CHICKEN	Hen	Rooster	Chick[1]	Capon[2] Stag[3]	107.5° (young: 102-106°)
DUCK	Duck	Drake	Duckling	n.a.	
GOOSE	Goose	Gander	Gosling	n.a.	
TURKEY	Hen	Tom	Poult	n.a.	
SHEEP	Ewe	Ram/ Buck	Lamb	Wether	100.9-103.8°
GOAT	Doe	Buck	Kid	Wether	101-102°
PIG	Gilt[6] Sow[7]	Boar	Shoat/ Piglet	Barrow[2] Stag[3]	101.6-103.6°
RABBIT	Doe	Buck	Kindle	n.a.	102.5°
VEAL CALF	Heifer	Bull Calf	Calf	Steer	98-102.5°

° This is an *average* period of *good* production. This is not to say a ewe will not give birth to a live lamb after the age of 12, or a goat will definitely not milk after the age of 12; rather, these are guidelines to help you determine how many productive years an animal will give you on the average. In the case of poultry, while they will lay after the age of two or three, their egg production will have decreased so markedly as to make them uneconomical, even for the backyard farmer.

Table of Helpful Facts

Age of Puberty	Incidence of Heat	Duration of Heat	Gestation	Avg. Productive Life*
4-6 mos.	n.a.	n.a.	21 days	2+ yrs.
5-7 mos.	n.a.	n.a.	28 days[4]	3+ yrs.
5-7 mos.	n.a.	n.a.	29-31 days[5]	3+ yrs.
n.a.	n.a.	n.a.	28 days	3+ yrs.
9 mos. #	13-19 days (avg. 16½)	3-72 hrs.	144-152 days (avg. 147)	10-12 yrs.
7 mos.	18-24 days (avg. 21)	1-2 days	145-155 days (avg. 150)	10-12 yrs.
3-5 mos.	16-24 days (avg. 21)	1-3 days	114 days	8-9 yrs.
6-7 mos.[8]	See text	See text	29-35 days (avg. 31)	3-4 yrs.
n.a.	n.a.	n.a.	n.a.	n.a.

See Text
1 Female: pullet; Male: cockerel
2 Castrated before sexual maturity
3 Castrated after sexual maturity
4 Muscovy 35 days
5 Canadian and Egyptian 35 days
6 Before having a litter
7 After having a litter
8 Medium-weight rabbit breeds

Appendix B.1.

Metabolic Disease	Poultry Affected	Cause
NUTRITIONAL ROUP (Lack of vit. A)	All.	Lack of vit. A.

Infectious Diseases		
AMYLOIDOSIS (Wooden Liver Disease)	Ducks and geese.	Unknown.
AVIAN LEUKOSIS COMPLEX (Fowl Paralysis or Lymphomatosis)	Chickens.	Viral infection.
BLACKHEAD	Turkeys (less frequently chickens).	A protozoan parasite.
BOTULISM	Geese, ducks.	Ingestion of bacteria that grows on decaying plant and animal matter.
BROODER PNEUMONIA	Chicks, poults, ducklings.	Fungus usually from moldy litter of feed.
COCCIDIOSIS	Chickens and turkeys; rarely: ducks.	A protozoan parasite.
DUCK PLAGUE (Duck Virus Enteritis)	Ducks, geese.	Passed in swimming water. Esp. in Eastern U.S. Also passed by migratory birds.

Common Poultry Diseases

Symptoms	Treatment	Prevention
Lameness; discharge from nostrils and eyes.	Give feed high in vit. A, cod liver oil, fresh greens, yellow corn, raw carrots.	Feed sufficient vit. A. Free Range.
Hardened liver.	None.	None.
Tumors on internal organs, paralysis, pale combs, weight loss.	None.	Buy stock from reputable poultrymen; cleanliness and proper feeding.
Droopiness, ruffled feathers, drowsiness, dark heads, yellowish diarrhea.	Check with vet.	Segregate chickens and turkeys; cleanliness.
Loss of control of neck muscles; may drown if swimming.	Epsom salts force-fed (by funnel). One tsp. dissolved in water per adult.	Cleanliness and removal of rotting material.
Listlessness, grasping, rapid breathing.	None.	Keep litter dry and brooder free from drafts.
Bloody diarrhea, listlessness, ruffled feathers, pale comb and wattles; off feed.	Check with vet.	Feeding medicated feed; keeping litter dry or raising on wire floors.
Watery diarrhea, droopiness and often death.	None.	Good sanitation, disposal of infected birds; keep from swimming water.

FOWL CHOLERA	All.	Bacterial infection.
FOWL POX	Chickens and turkeys.	Viral infection.
INFECTIOUS BRONCHITIS	Chickens.	Viral infection.
KEEL DISEASE	Ducklings.	Bacterial infection.
NEWCASTLE DISEASE	All.	Viral infection spread among infected poultry and on shoes, hands, carrying boxes, etc.
NEW DUCK DISEASE (Infectious Serositis)	Ducklings 1 to 5 weeks old.	Bacteria *Moraxella anatipestifer*.
PULLORUM	Chickens, turkeys (sometimes geese and ducks).	Bacterial infection. Present in adults and affects young before hatched.

Weakness, inactivity, ruffled feathers, off feed. Fever; head, comb, and wattles dark; greenish or yellowish diarrhea; mortality.	See vet.	Sanitation; no mudholes or stagnant water. Burn or bury infected birds.
Skin-type: small blisterlike spots on face, comb, and wattles. *Throat-type:* Yellow pus-like patches on mouth and throat.	None.	Vaccination if disease is present in your locale.
Gasping and coughing; sneezing. Markedly lower egg production; misshapen eggs.	None. Destroy flock and disinfect all houses and equipment before buying new stock.	Vaccination in infected locales.
Young ducklings thin and dehydrated.	None.	Clean incubator before using; reduce stress for young: provide warmth, water and feed. Good sanitation.
Similar to infectious bronchitis but increased water consumption and paralysis may occur.	None. Dispose of all stock and disinfect.	Strict sanitation; vaccination if present.
Sneezing, loss of balance. May fall on sides and back.	Antibiotics and sulfa drugs. See vet.	None.
White diarrhea, pasted-up vent. Adults: decreased hatchability of eggs and lower egg production.	Keep chicks warm and dry. See vet.	Buy chicks from certified Pullorum-free flocks.

VIRAL HEPATITIS	Duckling 1 to 5 weeks old.	Viral infection.
Parasitic Diseases:		
	While poultry are susceptible to a number of internal parasites, they are usually not enough of a problem to worry about. We have never wormed our flocks, but if greatly increased feed consumption and loss of weight is evidenced you can purchase a wormer that is mixed with drinking water. Lice can be a problem (more so for the chickens' or other poultry's comfort) and may if severe enough affect laying. Dust baths will control lice to a great extent.	

Appendix B.2.

Metabolic or Nutritional Disease	Cause	Symptoms
KETOSIS	Overfatness—lack of excercise. Reduced feed intake, large litter.	Just before or after kindling: listlessness, loss of appetite, diarrhea. Oversized liver, excessive fat in abdomen.
RICKETS	Calcium, phosphorus or vitamin D deficiency. Lack of sunlight.	Fore or hind legs crooked, bones fragile.
SLOBBERS	Excessive feeding of green feed. Indigestion. Abcessed molar teeth.	Wet about face, face may be swollen and cheeks may contain pus.

80%-90% mortality.	None.	Vaccination of infected flocks.

Common Rabbit Diseases

Treatment	Prevention	Metabolic or Nutritional Disease
None.	Don't overfeed does; encourage excercise. Provide palatable feed at kindling time.	**KETOSIS**
Supply calcium, phosphorus and vitamin D.	Feed balanced ration and supply direct sunshine.	**RICKETS**
Cut back on green feed. Remove abcessed tooth.	Limit feeding of green feeds.	**SLOBBERS**

Infectious Disease		
COCCIDIOSIS	A protozoan bacteria.	Listlessness, anemia, potbelly, thinness, loss of appetite, diarrhea.
ENTERITIS (NON-SPECIFIC)	Bacterial infection.	Scours, dirty behind. Feces dark, poorly formed and may stick to wire.
EYE INFECTION	Bacterial infection.	Babies may have eyes stuck shut, pus discharge, inflamed eyes.
MASTITIS (BLUE BREAST)	Bacterial infection; injury to mammary gland.	Swollen milk gland; tender, may be dark-colored and abcessed.
MUCCOID ENTERITIS	Unknown.	Most common 5 to 8 weeks. Excessive thirst, potbellied, diarrhea, feces watery or jellylike.
PASTEUREL-LOSIS-HEM-ORRHAGIC SEPTICEMIA	Bacterial infection coupled with stress.	Most common in fryers: listlessness, rapid breathing, potbellied and diarrhea.
PNEUMONIA	Bacterial or viral infection coupled with stress.	Rapid breathing, head held high, nasal discharge. Most common in does.
SORE HOCKS	Injury to foot followed by	Hunched up or lies about; painful

		Infectious Disease
Sulfa drugs. See vet.	Strict sanitation. Avoid manure buildup and prevent fecal contamination of feed and water.	**COCCIDIOSIS**
Medicated feed. See vet.	Prevent stress-wind, rain, poor housing. Avoid contact with infected stock.	**ENTERITIS (NON-SPECIFIC)**
Apply eye ointment daily.	Prevent cold drafts.	**EYE INFECTION**
Thoroughly clean and disinfect hutch and nest box. Administer 100,000 units of penicillin/10 lbs.	Cleanliness. Prevent injury to mammary gland on nest box or other objects in hutch.	**MASTITIS (BLUE BREAST)**
None.	Reduce stresses; avoid contact with infected stock.	**MUCCOID ENTERITIS**
Isolate; give greens and medicated feed.	Avoid stress; do not have contact with infected stock.	**PASTEUREL-LOSIS-HEM-ORRHAGIC SEPTICEMIA**
Isolate. Give 200,000 units penicillin, repeat in 72 hours.	Avoid stress and provide good sanitation.	**PNEUMONIA**
Clean affected area with soap	Have stock with thick pads. Avoid	**SORE HOCKS**

	infection. Insufficient floor support.	to walk. Scabs on bottom of hind feet.

Parasitic Disease		
EAR MITES	Ear mite.	Scabs in ears; scratching of ears. May lose weight.
FUNGUS INFECTION	Fungus infection.	Scaly skin on shoulders or along back; thin hair, dandruff.
RINGWORM	Specific fungus infection.	Loss of hair, skin inflamed in rings.

Miscellaneous Afflictions		
HEAT STROKE	Too much exposure to direct sun; lack of ventilation.	Panting, mouth open, quiet.
MALOCCLUSION (BUCK TEETH)	Inherited characteristic.	Buck teeth—lower teeth protrude, upper teeth long and curve into mouth.
VENT DISEASE	Infection by a spirochete, urine burn or hot metal floor burn.	Swollen genitals, blisters or dark scabs on same.

(Adapted from "Rabbits," a Carnation-Albers pamphlet)

and water and apply solution of Bluestone weekly. Supply soft bedding.	sharp objects or wire on floor of cage. Keep floor dry.	
		Parasitic Disease
Clean ear with cotton swab and apply weekly for 4 weeks: 5% phenol in sweet oil or 5% chlordane solution.	**Prevent contact with infected animals.**	**EAR MITES**
Apply 2% Lysol solution to infected area every other day for a week.	Prevent contact with infected animals.	**FUNGUS INFECTION**
Disinfect hutch, then dip infected animals in lime sulfur dip or see vet.	Prevent contact with infected animals.	**RINGWORM**
		Miscellaneous Afflictions
Submerge in cold water. Place in shade with adequate ventilation.	**Prevent direct exposure to sun and supply adequate ventilation.**	**HEAT STROKE**
Clip long teeth with sharp wire cutter.	**Breed from parents free from this trait.**	**MALOCCLUSION (BUCK TEETH)**
Clean off scabs and apply antibiotic ointment or powder.	Do not let moisture or manure build up in pen. Avoid contact with affected animals.	**VENT DISEASE**

Appendix B.3.

Metabolic or Nutritional Diseases	Cause	Symptoms
CONSTIPATION AND PINNING	Overeating; improper foods	Straining and distress in attempting to pass feces. Hard dry droppings. Pasted up anus.
SIMPLE SCOURS	Overeating; worm parasites; poor husbandry.	Diarrhea, watery feces, dehydration or shrinking in body weight.
BLOAT OR TYMPANY	Lush, young pasture, especially legumes; lack of roughage to excite belching.	Distension of rumen with excess gas; lack of appetite, increased breathing and heart action, death from suffocation.
IMPACTION OR ATONY	Broken-mouthed ewes on coarse hay; lack of vitamin A; heavy grain feed.	Poor appetite; poor condition generally; bad breath; lowered temperature; constipation, weakness and death.
MILK FEVER, HYPOCALCEMIA (LAMBING SICKNESS)	Lack of calcium.	Soon after lambing: loss of appetite, restlessness, failing, coma, death.

Sheep Diseases and Parasites

Treatment	Prevention	Metabolic or Nutritional Diseases
Warm soapy water enemas. 1-2 teaspoonsful of castor oil.	Do not overfeed; do not feed dry, coarse, indigestible feeds. Dock tail early.	**CONSTIPATION AND PINNING**
1-2 teaspoonsful of castor oil. Reduce feed intake. Do not excite.	Do not overfeed ewe or lamb; do not change feed too quickly; parasite control program.	**SIMPLE SCOURS**
Use mouth gag to promote belching; have veterinarian administer detergents; 1 teaspoonful aromatic spirits of ammonia in ½ pint cow's milk.	Feed hay before turning on lush legume pasture; use mixed pasture of grasses and legumes; keep off pastures until dew dries.	**BLOAT OR TYMPANY**
Good quality hay; fish liver oil; drench with water and knead flank to soften stomach contents; calcium gluconate intravenously; mineral oil or Epsom salts.	Proper nutrition; good pasture and hay; exercise.	**IMPACTION OR ATONY**
Keep warm; inject calcium salt intravenously.	Feed free choice at all times mineral mixture: ½ mineral salt; ½ dicalcium phosphate.	**MILK FEVER, HYPOCALCEMIA (LAMBING SICKNESS)**

PREGNANCY DISEASE, ACE-TONEMIA, KETOSIS	Lowered blood sugar; increased ketones in blood.	Occurs during last month of pregnancy; ewes carrying twins usually; muscular spasms, twitching of ears, loss of appetite; blind staggers, coma, labored breathing, frequent urination and death.
VITAMIN A DEFICIENCY OR NIGHT BLINDNESS	Low vitamin A content of liver.	Night blindness, occasionally sore eyes, some ulceration, loss of appetite, weakness, excitability and convulsions.
VITAMIN E DEFICIENCY, STIFF LAMB DISEASE, WHITE MUSCLE DISEASE	Deficiency of vitamin E; selenium.	Partial to generalized paralysis and stiffness; no pain; inability to nurse, lamb wants to nurse but can't get to udder. May starve to death.

Infectious Diseases		
DYSENTERY, INFECTIOUS	Bacteria of Colon group or Clostridium group.	Profuse diarrhea; maybe gray, yellow or blood tinged, fetid odor, lack of appetite, depression, exhaustion, sudden death.

Have veterinarian give glucose, molasses or Karo syrup if animal can swallow. Prevention best answer.	Avoid quick diet changes; feed ample balanced diet; include plenty of carbohydrates, feed at regular hours, give moderate amount of daily exercise.	**PREGNANCY DISEASE, ACETONEMIA, KETOSIS**
Place on green pasture, feed well-cured hay, fish liver oil if feeds not available.	Green feed and well-cured hay containing carotene. Feed corn which contains carotene. Feed Commercial Carotene carriers.	**VITAMIN A DEFICIENCY OR NIGHT BLINDNESS**
Have veterinarian inject pure vitamin E or tocopherols; concentrated cooked pressed wheat germ oil daily.	Adequate diet for ewe containing vitamin E. Mixed timothy and clover; grain 1 part bran, 2 parts oats; corn silage, wheat germ meal.	**VITAMIN E DEFICIENCY, STIFF LAMB DISEASE, WHITE MUSCLE DISEASE**

		Infectious Diseases
None—Prevention only presently known answer.	Strict husbandry hygiene; clean dry quarters; move pregnant ewes to clean quarters; green grass as early as possible; antitoxin used in England.	**DYSENTERY, INFECTIOUS**

ENTEROTOXEMIA OR OVEREATING DISEASE; PULPY KIDNEY DISEASE	*Clostridium welchii*, type D.	Lambs often found dead; convulsions; head may be held to one side, excitement followed by coma, high mortality; sugar in the urine; hemorrhages on intestine.
TETANUS OR LOCKJAW	*Clostridium tetani.*	Stiffness of legs, straddling gait, can't eat, tail rigid, sudden noises result in spasm, labored breathing, death. Sawhorse attitude.
SCRAPIE	Thought to be a virus.	Itching, incoordination, weakness, paralysis, and death. Incubation period may be up to 2 years; restless excitement; grinding of teeth; loss of wool from flanks; appetite good; no fever.
BLUE TONGUE	A virus.	Rise in temperature, lack of appetite, red to blue mucous membranes of mouth, some ulceration, offensive odor, discharge from eyes and nose, hair-hoof line becomes hot and tender,

None specific. Prevention is the only answer to date.	Excellent bacterin available. Do not force-feed heavy grains, take ewes off rich pastures.	**ENTEROTOXEMIA OR OVEREATING DISEASE; PULPY KIDNEY DISEASE**
No treatment very effective, some spontaneous recovery; prevention best answer.	Cleanliness in husbandry, cleanliness of instruments for docking, castrating, shearing; often follows use of rubber band castration; toxoid or anti-toxin sometimes used.	**TETANUS OR LOCKJAW**
Destruction of diseased animals only possibility now.	Since cause and transmission not known, no adequate control methods available. Vaccines not available. Slaughter and disinfect; federal indemnity and state indemnity in some states.	**SCRAPIE**
None. Prevention only answer to date.	Excellent vaccine now available in areas where endemic.	**BLUE TONGUE**

		resulting in lameness.
SORE MOUTH, CONTAGIOUS ECTHYMA	A virus.	Mouth, lips, nostrils of lambs, occasionally udder of ewe; blisters to pus to ulcers, scabs gray-brown in color, difficult to nurse and eat.
NAVEL ILL; JOINT ILL POLYARTHRITIS	Erysipelas, Corynebacteria, Staphylococci, a virus.	Rise in temperature, depression, loss of appetite, joints sensitive to pressure, may be swollen, weight loss, navel may be swollen.
LISTERIOSIS OR CIRCLING DISEASE	Listeria monocytogenes.	Animals sluggish, eat listlessly; may stand with mouth full of food, one ear may droop, blindness may occur, head drawn to one side; animal moves in circle; incoordination; high fever; coma and death.
Mastitis Diseases		
BLUE BAG, GARGET	Pasteurella spp., Streptococci, Staphylococci, Corynebacteria.	Straddling walk, lamb can't nurse, ewe may grind teeth from pain; udder swollen, red, painful, ewe often runs fever; milk yellow, flaky, thick.

Best thought now to leave alone; may soften scabs, paint with iodine.	Buy clean replacements	**SORE MOUTH, CONTAGIOUS ECTHYMA**
Antibiotics have proved useful; however expense involved means prevention best means of control.	Improved sanitary conditions at lambing time; use good antiseptic on navel stump, sterilize knife used in castrating or docking; get lambs on clean grass.	**NAVEL ILL; JOINT ILL POLYARTHRITIS**
Nearly always fatal, antibiotics destroy organisms, but tissue damage usually irreparable.	Since disease found in rodents, take steps to exterminate them; isolate sick animals.	**LISTERIOSIS OR CIRCLING DISEASE**

		Mastitis Diseases
Remove lamb; milk gently by hand frequently, hot Epsom salt packs 3-4 times daily, antibiotics systemically or locally.	Tag out wool from udder, clean bedding, clean up discharges from uterus, remove causes of bruises such as high doorsills.	**BLUE BAG, GARGET**

FOOT ROT, INFECTIOUS PODODERMA-TITIS	Necrophorous organism and/or spirochetae.	Lameness in one or more legs; red swollen hoof-hair line; loosened hoof wall; foul-smelling discharge; loss of weight.
Parasitic Diseases		
Internal		
TAPE WORM	*Moniezia expansa.*	Diarrhea, loss of weight and in some cases acute convulsions ending in death. Anemia sometimes noted.
STOMACH WORMS	Wire worms, brown hair worms, bankrupt worms.	Anemia due to hemorrhage, loss of weight, harsh hair coat, dullness, listless, bottle jaw or swelling in the throat region.
SHEEP BOT (Grub in the Head)	Larvae of the fly *Oestrus ovis.*	Sheep on pasture hysterical in efforts in preventing flies laying eggs in nostrils, shake heads, stamp feet, bunch together holding noses to ground, bury nostrils in wool of other

Trim feet of all loose and dead tissue, bath in copper sulfate solution twice daily (2½ lbs. per gallon).	Pare feet regularly, keep sheep out of wet, swampy areas; keep barn and bedding dry.	**FOOT ROT, INFECTIOUS PODODERMATITIS**
		Parasitic Diseases
		Internal
See text.	Eggs carried in many species of pasture mites. Pasture remains infected at least 2 years after infected sheep leave; therefore, must treat sheep when infected.	**TAPE WORM**
See text.	Since these worms have a direct life cycle, rotating pastures every 2-3 weeks keeping a phenothiazine salt mixture on pasture continuously will control, balanced rations will prevent.	**STOMACH WORMS**
Nothing effective.	If only few sheep, keep in dark stables during day when flies are about.	**SHEEP BOT (Grub in the Head)**

		sheep, sneezing, coughing, nasal discharge.
INTESTINAL WORMS	Trichostrongyles, Cooper's Worm, Hook Worm, Nodular Worm.	Anemia due to hemorrhage, loss of weight, harsh hair coat, dullness, listless, bottle jaw or swelling in the throat region.
LUNG WORMS	*Dictyocaulus filaria.*	Coughing, pneumonia, sometimes death.
External		
LICE, TICKS (KEDS)		Scratching and rubbing; unthrifty, slow-gaining animals.

Source: Adapted from "The Production and Marketing of Sheep in New England."

Appendix B.4.

Metabolic or Nutritional Diseases	Cause	Symptoms
BLOAT		
KETOSIS		
SIMPLE SCOURS	See Appendix on Sheep Disease.	
MILK FEVER	Lack of Calcium.	Soon after kidding: loss of appetite, restlessness, falling, coma, death.

See text.	Since these worms have a direct life cycle, roating pastures every 3 weeks keeping a phenothiazine salt mixture on pasture continuously will control, balanced rations will prevent.	**INTESTINAL WORMS**
See text.	Rotate pastures, isolate infected animals; good sanitation.	**LUNG WORMS**
		External
See text.		**LICE, TICKS (KEDS)**

Common Goat Diseases

Treatment	Prevention	Metabolic or Nutritional Diseases
		BLOAT
		KETOSIS
		SIMPLE SCOURS
Keep warm; inject calcium salt intravenously.	Feed free choice at all times mineral mixture: ½ mineral salt ½ dicalcium phosphate	**MILK FEVER**

Infectious Diseases		
COCCIDIOSIS	Protozoan parasite in intestinal tract.	Lack of appetite, loss of weight, bloody diarrhea.
ENTEROTOX-EMIA FOOT ROT	See Appendix on Sheep Disease.	

Parasitic Diseases	
WORMS	See Appendix on Sheep Disease.
LICE	As with sheep, they can be a problem. Goats that are infested with lice can be observed scratching and rubbing frequently. Lice can also be detected by parting the hairs and inspecting the skin of the goat. For treatment apply a commercially available product containing *Coumaphos* as directed. The watering can method, as with sheep, can be used. *CAUTION:* Take great care to avoid getting this or any insecticide in milk or on milking utensils. Follow directions carefully for use with lactating animals.

NOTE: Read carefully directions of all medications for goats, specifically as they relate to milking animals. If in doubt, do not use any milk from the goat until medications have safely cleared their system.

Appendix B.5

Metabolic or Nutritional Diseases	Cause	Symptoms
ANEMIA	Lack of copper and iron.	Loss of appetite, weakness, swollen around head and shoulders. May have trouble breathing.

		Infectious Diseases
Sulfa drugs and antibiotics. See vet.	Strict sanitation; plenty of sunlight.	COCCIDIOSIS
		ENTEROTOX-EMIA
		FOOT ROT

		Parasitic Diseases
		WORMS
		LICE

Common Pig Diseases

Treatment	Prevention	Metabolic or Nutritional Diseases
See prevention.	1 to 1½ cc. iron dextran injection at 2 to 3 days of age.	ANEMIA

Infectious Disease		
BORDETELLA RHINITIS (ATROPHIC)	Bacteria which enters wounds in the mouth or nose. Spreads among pigs through food, water and body contact.	Young pigs: sneezing; 4-10 weeks snout wrinkles and may become distorted. Discharge from nose.
BRUCELLOSIS (Bang's Disease)	Bacteria spread through feed, water, body contact and afterbirth.	Hard to spot beforehand. Often abortion, sterility, and inflammation of joints and uterus or testicles.
COLIFORM ENTERITIS (Necro)	*E. Coli* bacteria.	Fever, loss of appetite, scours. Temperature often drops in a few days, and they begin to eat, but they gain poorly and mortality is high.
ERYSIPELAS	Microorganism passed through feces and urine and from feed and water used by infected pigs.	High fever, loss of appetite, immobility. Red, diamond-shaped blotches that turn white when pressed.
FLU	Virus passed among hogs when resistance is low.	Off of feed, listless, coughing and discharge from eyes. Fever may be present for a few days.
HOG CHOLERA	Not as prevalent now. Highly infectious: spread by urine, feces,	Fever, loss of appetite; eyes have sticky discharge and under-

		Infectious Disease
None.	Get rid of sows that have infected litters. Adequate calcium in ration. Well-fed pigs are less likely to be infected.	**BORDETELLA RHINITIS (ATROPHIC)**
None.	Check for infection by blood test; isolate infected animals.	**BRUCELLOSIS (Bang's Disease)**
Isolation; drugs as recommended by veterinarian.	Strict sanitation. Practice McLean County System of Sanitation. Feed sow and litter adequately.	**COLIFORM ENTERITIS (Necro)**
Isolation; treatment with anti-erypsipelas serum and penicillin.	Vaccination if disease is in your area.	**ERYSIPELAS**
Will run its course in a week. Supply plenty of water.	Eliminate drafts in pens and rapid temperature changes.	**FLU**
Contact authorities; destroy animal.	Vaccination.	**HOG CHOLERA**

	mouth and nose secretions, trucks, birds and streams.	side of neck and abdomen may have dark red or purple coloration. Cough.
LEPTOSPIROSIS	*Leptospira* in kidney or urinary tract.	Similar to erysipelas and hog cholera. May cause abortions and poor milk flow.
MMA SYNDROME (Metritis-Mastitis-Agalactia)	Transmitted by boars at breeding or contamination of birth canal with manure at farrowing. Lack of exercise during pregnancy may also contribute.	Sow may lie on belly and may not allow piglets to nurse. Udder may be hard, swollen and hot. Whitish or yellowish vaginal discharge. Off feed, fever. There is little milk and many piglets may die.
GASTROENTERITIS	Virus spread by direct or indirect contact.	Vomiting and diarrhea that resembles curdled milk. Acute in older pigs; terminal for young.
VIBRIONIC DYSENTERY	Bacteria in feces of infected animals.	Bloody diarrhea and black feces. More serious in young pigs.
VIRUS PIG PNEUMONIA (VVP)	Viral infection of respiratory tract.	Sneezing and coughing; off feed and retarded growth. Most serious in young pigs.

Terramyacin fed for 14 days is effective in control.	Vaccination.	**LEPTOSPIROSIS**
Contact vet. A systematic antibiotic/cortisone/hormone treatment will check infection and may restart milk flow. Feed piglets whole cow's milk (Jersey milk is best) every three hours.	Cleanliness at breeding and farrowing. Exercise.	**MMA SYNDROME** (Metritis-Mastitis-Agalactia)
None.	Vaccination; plenty of sunlight.	**GASTROENTERITIS**
Isolation; move to clean quarters. Treatment as described by vet.	Good sanitation.	**VIBRIONIC DYSENTERY**
Tylosin and antibiotics to control secondary infections.	Keep warm and free from chills.	**VIRUS PIG PNEUMONIA (VVP)**

Parasitic Diseases		
Internal		
LUNG-WORMS	Worms ingested while rooting.	Weak and coughing pigs; slower growth.
ROUNDWORMS (Ascarids)	Swallowing eggs in feed or water.	Weak, slow weight gains; pigs less resistant to disease.
TRICHINOSIS	Larvae of the parasite *Trichinella spiralis* embedded in muscle tissue. Passed to pigs by eating infected pork or from droppings of infected rodents or other furbearing animals.	None. Detectable only by microscopic examination.
External		
MANGE	Mites that feed on tissues of skin and blood.	Itching. Infection begins around eyes and ears; red rash and scaly hide.
LICE	Lice, grayish-brown in color.	Itching. Appearance of lice. Most often in winter.

		Parasitic Diseases
		Internal
Good commercial wormers are available in feed stores.	Rotate pastures; practice McLean County Sanitation.	**LUNG-WORMS**
		ROUNDWORMS (Ascarids)
No treatment. *Prevention in Meat:* Cook all pork to 137°. Larvae will also be killed by thorough curing or freezing at 5° or below for 3 weeks.	Keep garbage and feed in covered containers away from rodents. Boil questionable garbage and feed only what will be cleaned up at one feeding to prevent rat infestation.	**TRICHINOSIS**
		External
Spraying with mite-killing solution. Also, see below.	Sanitation.	**MANGE**
Oil bath with nondetergent crankcase oil. A pig oiler: a pole covered with burlap soaked in oil for pigs to rub on.	Sanitation.	**LICE**

Appendix B.6.

Infectious Diseases	Cause	Symptoms
COCCIDIOSIS	Bacteria, *coccidia*, getting into intestinal tract in large numbers.	Bloody diarrhea, loss of appetite, weakness.
FOOT ROT	Microbe entering feet through cuts or bruises.	Lameness; off of feet because can't move too easily.
PNEUMONIA	Often secondary to other infections. Chills or damp weather. Most common 6-8 weeks of age.	Temperature 104°-106° F. Labored breathing, coughing, weakness, loss of appetite. Wheezing or gurgling.
SHIPPING FEVER	Viral infection. Occurs most often when resistance is broken by shipping in cold, wet conditions and when animals are overcrowded and feed regularity is disturbed.	High fever (104°-107°F.). Cough, swollen eyes, watery discharge from nose.

Parasitic Disease		
STOMACH WORMS	Feeding or watering from dishes used by infected stock. Housing in contaminated pens.	Loss of weight, weakness, anemia (paleness of skin around mouth and eyes, swollen, "bottle jaw").

Common Calf Diseases

Treatment	Prevention	Infectious Disease
Often successful with sulfa drugs. See vet.	Overall cleanliness; clean pen daily. Clean water and feed dishes daily.	**COCCIDIOSIS**
Sulfa drugs. In severe cases, trimming rotted part and treating with a 10%-30% copper sulfate solution or a 2%-10% formaldehyde solution.	Keep barn floor dry and free from any objects that might injure feet.	**FOOT ROT**
Move to warm, dry place. Antibiotic treatment.	Keep dry and free from chills.	**PNEUMONIA**
Antibiotics.	Proper shipping conditions.	**SHIPPING FEVER**

		Parasitic Disease
Worming capsules available in feed stores, by mail from veterinary supply companies (see *Appendix 242*) or from vet.	Cleanliness	**STOMACH WORMS**

Appendix C. A Homemade Incubator

While you can buy an incubator for hatching eggs, a homemade one is so simple to make that it's silly not to go that route. The only thing you have to purchase is a thermostat, and possibly three lamp sockets and a few feet of wire.

Construction.

Simply build a wooden box, either of plywood or scrap lumber, of the dimensions indicated. Drill ½-inch vent holes at a height of 1 inch above the egg drawer in each of the four sides (including the door) and two holes in the top. The egg drawer should be built as indicated and slid into the incubator between the support on each side so that it can be pulled in or out to check the eggs and turn them.

The window in front will enable those interested to watch the hatching process without constantly opening the door and lowering the inside temperature. The glass can be taped on over the

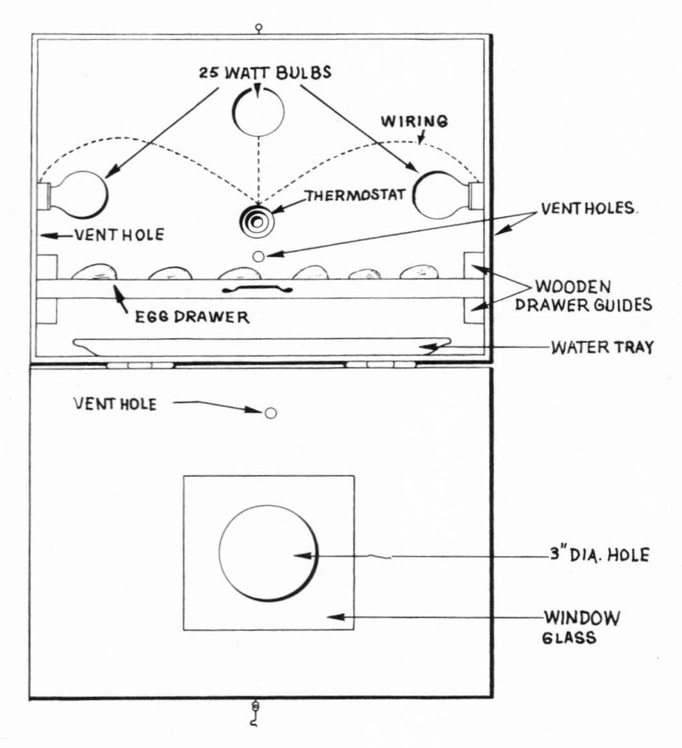

hole or glued on, using a good epoxy. Hinge the door and use a hook and eye or other fastener to insure that the door closes securely. The three light sockets should be mounted about seven inches from the floor of the incubator and wired into the thermostat. Although I have not tried it with an incubator of this design, you might try wiring one of the sockets so that the light is constantly on. Then you will be able to watch the entire hatching process without having to worry about the theromstat cutting the lights off at the most exciting times.

The one bulb left on all the time should not keep the incubator too hot. If it does, replace it with a 15-watt bulb, and the incubator should operate within the desired temperature range.

The thermostat should be installed two to three inches from the level of the eggs in the egg drawer. It will then most accurately reflect the temperature that affects the eggs. A good unit for use in a homemade incubator can be purchased through a farm supply catalog or local hardware or feed store. The vent holes drilled into the sides and top of the incubator will insure proper ventilation.

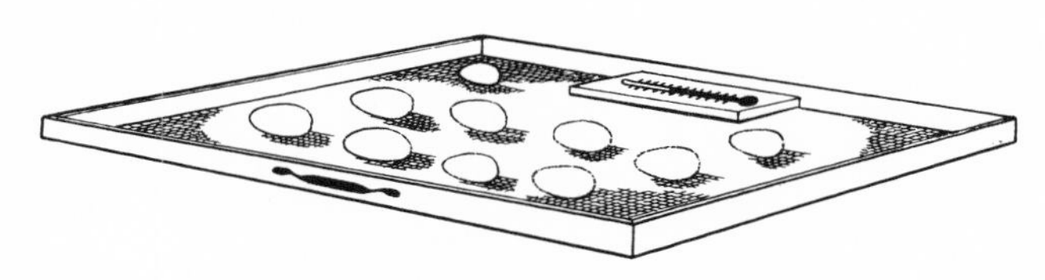

Operation.

Turn your incubator on the day before you are ready to begin hatching your eggs so that it will heat up and set the thermostat at the proper temperature. Many thermostats don't have any calibrations, and you'll have to find the correct temperature set by trial and error. Remember, you'll want to set the temperature within 99°-103° F. The temperature should remain between these extremes. Warm your eggs up to room temperature before placing them in the incubator so the temperature will not fall sharply when they are put in.

The pan of water is very important as it will help to maintain the proper humidity in the incubator. Keep it full at all times and do not permit the water to get stagnant. Finally, it is essential to turn the eggs four times a day, at eight-hour intervals if possible. Make a *small* mark on one side of the egg to avoid confusion when turning.

TABLE 1: Average Hatching Period for Common Fowl

Chickens	21 days
Geese	
Canadian and Egyptian	35 days
All other breeds	29 to 31 days
Ducks	
Muscovy	35 days
All other breeds	28 days
Turkeys	28 days

Appendix D. Giving an Injection

As you acquire livestock, there will be times when you have to administer medicine by injection. To have a veterinarian travel out to your house every time a shot is needed would be prohibitively expensive, and I doubt whether a veterinarian would even make a special trip just to give one. The answer is to learn to give your own shots, and this is not nearly so difficult—or dangerous—as you think. Once you learn how to administer your own shots, you can order most of your pharmaceuticals from veterinary supply houses (listed in another appendix) and realize even more substantial savings. An example: for our first litter of pigs we purchased 1 cc. of iron for each of six pigs from a local vet for a total of $6. I have since ordered my own iron shots by mail at a cost of less than $2 for 20 cc.

I am certainly not advocating that you begin, hog-wild, prescribing your own medicines without the advice of a vet and ordering them by mail. But you can order such commonly used items as iron shots for pigs, penicillin, tetanus toxoid and antitoxin and other drugs recommended for use by a veterinarian. I have also phoned in an order to a company for a drug a vet has prescribed and received it in the mail two days later.

Remember to follow carefully all directions in administering, storing, and dosage of drugs.

Preparation

Boil all needles and syringes for ten minutes before use or thoroughly disinfect with alcohol. Clipping the hair around the site of the injection may be necessary for certain shots. To fill a syringe, turn the bottle upside down and insert the needle into the rubber stopper and into the bottle. Draw the plunger of the syringe down until the correct amount of serum is received. I usually have the plunger halfway withdrawn when I insert the needle in the bottle, then I push the plunger in, thereby creating pressure in the bottle and forcing the serum out more quickly. Be sure the needle touches nothing after it is filled and before it is used.

There are four basic types of injection: intramuscular, subcutaneous, intradermal and intravenous. Intravenous injections are not used for any of the common shots we might give (iron, penicillin, tetanus, etc.) and involve more problems (location of vein, danger of air in veins, etc.) and it is best to consult a veterinarian if one is needed.

Intramuscular

This is simply the injection of a substance into the muscle tissue. In poultry, choose a site in the fleshy part of the thigh. For other animals choose a site either in the thigh or in the muscle of the neck. Smaller animals can be restrained by a helper but larger animals, notably pigs, can present some problems. Sheep and goats can be seated on their rumps or backs as in a shearing position and this should quiet them. A small calf can be restrained by hand or be tied if it is larger. With pigs, I give them a tempting bit of food and inject the serum in the thigh. Often the fright of being poked suddenly is worse to the animal than the actual prick of the needle. To prevent this, tap the area first with your finger a few times then insert the needle. If the animal jumps around and shakes the needle off before you have time to inject it, try removing the needle from the syringe and inserting that first. After the animal has settled down, attach the syringe and administer the shot.

In an intramuscular shot, insert the needle at right angles to the body and withdraw the plunger a bit before injecting. If any blood is drawn into the syringe you have hit a blood vessel and you should choose another site before injecting.

Subcutaneous

This is injecting a substance under the skin. Choose a site where the skin is loose: the breast of poultry, behind the ear of pigs, the chest or neck of a calf or rabbit, and the chest or inner thigh of sheep or goats. Pinch and lift up a fold of skin and insert the needle through the skin and beneath it. Before injecting, withdraw the plunger a bit to see if any blood is drawn up. If not, inject the dose; if so, find a new site.

Intradermal

This is a shot within the layers of skin rather than beneath it. Choosing a site as with subcutaneous injections, pinch the skin between the thumb and forefinger so as to make a mound of skin a few inches long. Using a thin needle, insert it parallel to the body and just under the surface of the skin. Inject the dose as you are slowly withdrawing the needle and place a finger over the hole if any begins to leak out.

Appendix E. Tanning Skins

It is an unpardonable waste, in my opinion, to throw the skins of your animals into the trash. If that is not enough, perhaps you will be convinced by the fact that you can make lovely and useful hats, gloves, blankets, etc., from your tanned hides and even sell and barter them for additional income or goods. Sheepskins and rabbit skins first come to mind, but don't neglect calf skins or even your goat hides. The industrious person will find uses for all these skins. Rabbit skins are perhaps the easiest to tan because of their size and because there is relatively little fat. Sheepskins are by far the most difficult (but you get the best finished product) because of their size and because of the large amount of fat that must be removed from the skin. In fact, many taxidermists won't even consider tanning a sheepskin because of all the bother they can be. But that's because they scrape the fat off by hand, and that is a monumental job, believe me, because I've spent a day or two at it. It's a long and hard job and does not give very satisfactory results. The following methods will enable you to tan any type of skin (including sheepskins!) with a minimum of time and effort.

Preparation of Skins

It is best to use skins from freshly butchered animals but those that have been dried or salted can be used as long as they are soaked well and are pliable before tanning. Wash skins in lukewarm water with a little detergent to remove any loose dirt and bloodstains. Rinse well in cold water and wash as often as is necessary to remove all blood.

Fleshing

This is why tanning has never caught on—fleshing *can* be terrible. With calf skins, rabbit skins, and other animal skins that don't have too much fat, you can do the fleshing by hand in a few minutes and go right on to the next step, tanning. With sheep and goat skins, this fleshing, or the removal of meat and fat from the skin, is a substantial job without the use of the pickle bath mentioned below. The following fleshing/tanning bath is supplied by Jerome Belanger in his book *The Homesteader's Guide to Raising Small Livestock:*

Mix: 8 fluid oz. battery acid or
2 fluid oz. sulfuric acid
2 lbs. (3 cups) pickling salt
and 2 gallons water

While you might be leery of handling sulfuric acid, you can use the battery acid. It is not as dangerous, is quite cheap and readily available in auto supply stores. Mix enough of the solution to cover the skin and place it and the skin in large stoneware crock or plastic container (don't put any acid in a metal container or it will corrode it). Weight the skin down so it won't rise out of the solution. A small hide may be done in less than a week, while a large sheepskin may take up to two weeks. You will learn how long with experience, but don't worry in the meantime about damaging the pelts by keeping them in too long. They can stay in almost indefinitely as long as they are stirred about in the solution periodically.

If the container is small and the skins are tightly packed, take them out every day and rearrange them so all sides get exposed to the tanning solution. In a larger container, or with smaller skins, take them out every three days, stir the mixture and reimmerse them. The temperature of the solution is quite critical—keep it as close to 70° F. as possible. I tanned some skins one fall in our upstairs; the temperature fell into the fifties up there for a week and the skins spoiled. Apparently it was too cool for the tanning to take place, but unfortunately not cool enough to retard the rotting process. The odor is minimal with this process, so you can keep it in your living area if the temperature elsewhere is too cool.

When finished, the hide should be taken from the solution and rinsed in cold water to remove any remnants of the acid. If all goes well, the flesh and fat will peel off the hide in large sheets.

If you have ever fleshed a hide by hand you will experience almost orgasmic joy at how much easier this process is. What will take hours upon hours by hand will take perhaps 10 or 15 minutes after being in the acid bath. If the fleshing is still hard to complete, place it back in the solution for a few days until the fat peels off easily. Once the fleshing is complete, rinse, and put back into the solution for another week. After this, wash in warm water and rinse well. I have gotten good results using this method alone, but if your results are not satisfactory (in regard to the quality of the leather and hide after tanning), you can go on to the tanning procedures, listed next. Again, I have found these steps unnecessary, so if you try it and agree, skip the next section and go right on to finishing.

Tanning

After fleshing, either by hand or with the above process, you can tan using one of the following methods:

Alum Mixture

Mix: 1 lb. alum
2 lbs. (3 cups) salt
Mix with enough water to cover skins

Dissolve the alum first in hot water, then add salt and enough cold water to achieve the correct solution. Keep the skin in this solution for two days and turn three times a day. (Alum, an astringent, can be purchased in most pharmacies.)

Tannol Mixture

Mix: 1 lb. Tannol
2½ lbs. (3¾ cups) salt
10 gallons water

First use an acid bath as explained above. Keep hide in this solution for five days and stir three times a day. (Tannol is an aluminum sulfate compound which can be purchased from the Northwestern Fur Company, Omaha, Nebraska.

Finishing

If you want to remove the hair from your hide, you can usually do so by pulling it out after it's been either in the acid bath or in one of the tanning solutions (the hide must still be wet for the hair to come out easily).

After removing the hide from whichever solutions you use, rinse the skin in cold water to remove any solution that may be left. Then let the skin drip dry out of sunlight. Before the skin is completely dry and is still pliable the leather side may be oiled. Then you must work the leather to make the hard, brown skin soft and white. Constant stretching, pulling and rubbing over a smooth board, will do the trick in time. Be sure to do this work before the skin dries out. Another easy way to finish the hide is first to stretch it and nail it to a frame (any holes or cuts in the hide should be sewn up before stretching to prevent further ripping of the hide), and then sand it with an electric sander or by hand. After the leather is softened comb burrs and other chaff out of the fleece with a metal comb, trim the skin to even it up (I usually cut the legs off a hide), and you're all set.

Appendix F.

Animal	Weight (1)	Per 1000# Per Year (Tons)	Manure (2) Per Animal Per Day (3) (Lbs.)	Per Animal Per Year (Lbs.) (3)
CHICKEN	5	9.6	.26	95
SHEEP AND GOAT	100	7.3	4.0	1,460
PIG	500	11.8	32.5	11,860
RABBIT	4	4.2	.09	34
VEAL CALF	100	10.9	6.0	2,190

(1) This is an average figure. For chickens, sheep and goats, and pigs this is an average for an adult. For the rest of the animals it is a point ¾ between birth weight and slaughtering weight.

(2) Raw manure. That is, urine and feces with no bedding.

(3) Based on average weight from Column 1.

Manure Table

% Urine	% Feces	% Nitrogen	% Phosphorus	% Potassium
n.a.	n.a.	1.4	0.5	0.6
50	50	1.1	0.2	0.8
45	55	0.7	0.2	0.5
50	50	2.4	1.4	0.6
50	50	0.6	0.2	0.4

NOTE: There can be up to a 20 percent variation in the amount of manure depending on feed, environment and health factors. Urine, pound for pound, is of greater fertilizing value than feces (except in pigs), so good absorbent bedding is most important. Storage is also important. Manure that is left outside will have less value than that which is under some cover.

Remember also that much of the manure will be voided on pasture. To determine how much you can save, figure the number of days the animal is confined and multiply times the entry under pounds of manure per day. Similarly, for those animals that don't live a full year, figure the number of days alive and multiply times pounds per day.

Appendix G.
A Backyarder's Catalog of Resources

Here is a list of useful books, periodicals, catalogs and other sources of information, most of which are related directly to the text of *Backyard Livestock*. Other references range farther afield into the whole subject of country living, self-sufficiency, preserving meat and vegetables, marketing wool, etc. Suppliers of equipment, veterinary products, and even some kinds of livestock are included. Book prices are for hardcover editions unless otherwise noted. The names and addresses of publishers cited more than once appear at the end.

USDA Publications Free publications on a variety of topics. Write for free catalog: "List of Available Publications," U.S. Department of Agriculture, Washington, D.C. 20242.

and/or "Popular Publications for the Farmer, Surburbanite, Homemaker, Consumer," U.S. Department of Agriculture, Washington, D.C. 20242.

State Extension Services

Your local country extension agent will gladly offer free advice on raising livestock, growing crops, and the like. Many excellent publications are offered free to state residents and at a nominal cost to out-of-staters. Your local extension agent usually has an office in the county seat. If you can't locate him write:

Extension Service Director
c/o College of Agriculture
Your State University

GENERAL

Books

Bundy, et al., Livestock and Poultry Production, 4th ed., 1975. Prentice-Hall ($16.50). (Technical, but a good book on poultry, pigs, cattle and dairy cows, sheep and horses.)

Ensminger, M.E., The Stockman's Handbook, 1970. Interstate Publishers ($16.50). (Technical, but crammed with information, especially on feed and diseases. Covers beef, pigs, sheep and horses.)

Herriot, James, All Creatures Great and Small, 1973. Bantam Books, 666 Fifth Avenue, New York 10019. (paperback, $1.75); All Things Bright and Beautiful, 1975. Bantam Books (paperback, $1.95). (These are great books. And who knows, you may find yourself with your hand in a sheep's vagina, too.)

Kains, M.G., Five Acres and Independence, 1975 (orig. 1935). Dover Publications (paperback, $2.50). (Good all-around manual on small farm management.)

Langer, Richard, Grow It!, 1972. Avon Books (paperback, $3.95). (Animals, crops, farming, etc. All about making it on the land.)

Seymour, John and Sally, Farming For Self-sufficiency, 1973. Schocken Books ($7.50).

Wiggington, Eliot, ed., The Foxfire Books (I, II and III). Doubleday Publishing Co. (Excellent books on heretofore lost arts dealing with farming, cooking and self-sufficiency.)

Periodicals

Country Gentleman, 1100 Waterway Blvd., Indianapolis, Indiana 46202. Quarterly. Subscription $6.00/year. (The oldest agricultural journal in the world. Articles on livestock, gardening and farming.)

Countryside and Small Stock Journal, Rte. 1, Box 239, Waterloo, Wisconsin 53594. Monthly. $9/year. (Gardening, small stock: poultry, rabbits, sheep, goats. Excellent. Best all-around magazine of its type. A must.)

Country Journal, 139 Main St., Brattleboro, Vermont 05301. Monthly. $10/year. (Nice glossy country publication; entertaining and informative.)

Farmstead Magazine—Maine Gardening and Small Farming, Box 392, Blue Hill, Maine 04614. Quarterly plus annual. $4/year. (For Maine climate but has wider application.)

Landward Ho!, Wilkinson Press, Rte. 1, Scio, Oregon 97374. Monthly. $4/year. (Mimeographed magazine stressing old, forgotten recipes and methods.)

Mother Earth News, Box 70, Hendersonville, N.C. 28739. Bimonthly. $10/year. (A must; the best all-around back-to-the-earth publication. Crammed with information on livestock, gardening and farming, alternative energy, etc.)

Organic Gardening and Farming Magazine, Emmaus, Pa. 18049. Monthly. $6.85/year.

Correspondence Courses

"Correspondence Courses," Pennsylvania State University, 307 Agricultural Administration Building, University Park, Pa. 16802. Free booklet. (Many correspondence courses on livestock and other aspects of country living. From $3 per course.)

Miscellaneous

Earth Move, PO Box 252, Winchester, Mass. 01890. (An organization dealing with alternative energy, farming, self-sufficient living. They are planning to put out, sometime in 1976, a magazine *The New Earth Times*, "a radical self-sufficiency quarterly." Among other things, it will deal with "farming and growing." $10/year for magazine and a "New Earth Catalog" supplement. Sounds intriguing.)

The New Alchemists, PO Box 432, Woods Hole, Mass. 02543. (Conducting ongoing research into alternative energy, land use and agriculture. Write for information on their publications.)

Catalogs

Countryside General Store, 312 Portland Road, Waterloo, Wisconsin 53594. (Stoves, butterchurns, butchering tools and other equipment for country living.)

Cumberland General Store, Rte. 3, Box 479, Crossville, Tenn. 38555. ($3 for catalog, refunded on first order. All types of hard-to-find and previously out of production tools, feed mills, etc., etc.)

C.H. Dana Co., Inc., Hyde Park, Vermont 05655.

Metastasis, Box 128-C, Marblemount, Washington 98267. (13¢ stamp for catalog of hard-to-find information, tools and supplies.)

Montgomery Ward's Farm Catalog, 618 West Chicago, Chicago, Ill. 60610. (Tools, fencing, supplies.)

Mother's General Store, Box 506, Flat Rock, N.C. 28731. (Sells almost anything you need for country living as well as livestock supplies.)

Nasco Farm and Ranch Catalog, Nasco, Fort Atkinson, Wisconsin 53538. (The most complete catalog on livestock supplies. Has everything but pharmaceuticals. If they don't have it, I don't know who does.)

Sears Farm Catalog, 313 E. Ohio St., Chicago, Ill. 60611 or from your local catalog store. (Good range of supplies at reasonable prices. Fencing, tools, incubators, brooders, day-old chicks, ducklings, goslings, guinea hens, etc.)

Whole Earth Catalogs, Epilogs and other spinoffs. (Keep you abreast of what's new and how to gain access to almost anything.)

Book Catalogs

Books for Garden Way Living, Garden Way Publishing Company. (Free catalog on scores of books for country living.)

Diamond Farm Book Publishing, Dept. SJ, Rte. 3, Brighton, Ontario, Canada KOK 1HO. (Free livestock book list.)

Interstate Publishers. (Since so many of the books in this section are published by this company it would be worthwhile to send for their current list.)

Mother's Bookshelf, Box 70, Hendersonville, N.C. 28739. (Free catalog of books on country and self-sufficient living.)

POULTRY

Books

Klein, G.T., Starting Right with Turkeys, Garden Way Publishing (paperback, $3).

Mercia, Leonard, Raising Poultry the Modern Way, 1975. Garden Way Publishing (paperback, $4.95). (Detailed; covers chickens, ducks, geese and turkeys.)

Smith, Page and Daniels, Charles, The Chicken Book, 1975. Little Brown and Co. (Not really a how-to-book, but goes into detail on the history, rise and fall of the chicken. An enjoyable book if you love chickens—or even if you hate them.)

Poultry Handbook, Penn. State University (paperback, $1.50).

Poultry Health Handbook, Penn. State University (paperback, $2.50).

Equipment

Stromberg's, Box 717, Ft. Dodge, Iowa 50501. (Equipment, books, many, many breeds of poultry available. You can even order monkeys and a buffalo!)

Brower Manufacturing Company, Box 7522, Quincy, Ill. 62301. (Brooders; free catalog.)

See also "General Catalogs."

Hatcheries

Breakfast Hill Hatchery, Greenland, N. H. 03840. (Chickens.)

Murray McMurray Hatchery, M171, Webster City, Iowa 50595. (Listings, 25¢.)

Stromberg's (above.)

Willow Hill Farm and Hatchery, Dept. B, Richland, Pa. 17087.

See also advertisements in periodicals and consult local telephone directories.

RABBITS

Books

Templeton, George S., Domestic Rabbit Production, 1968. Interstate Publishers ($7.95). (The best.)

Periodicals

National Rabbit Raiser, 241 W. Snelling Ave., Appleton, Minnesota 56208.

The Rabbitman, Auburn, Alabama 36830.

Equipment

Favorite Manufacturing, Inc., RD 1, Box 176, New Holland, Pa. 17557. (Cages, equipment, watering devices, scales, etc.) Free literature.

Glick Manufacturing, Box 1649, Gilroy, California 95020. ($1 for catalog and guide to raising rabbits. Refundable on orders of $10 or more.)

Morgan Enterprises, Box 316, Liberal, Kansas 67901. (Catalog, 50¢, refunded on first order.)

See also "General Catalogs."

Associations

The American Rabbit Breeders Association, 4323 Murray Ave., Pittsburgh, Pa. 19115.

EARTHWORMS

If you're going to raise rabbits (or any livestock for that matter), you you should really consider growing earthworms (in their manure) for your own use and profit.

Earthworms for Ecology and Profit. Earthworm Oasis, Inc., 40892 Harper Lake Rd., Hinkley, Calif. 92347 ($5). (Send for free information on worm raising plus prices for earthworms.)

Shields, Earl B., Raising Earthworms for Profit. Garden Way Publishing (paperback, $2.50).

SHEEP

Books—General

Cooper, M. McG., and Thomas, R.S., Profitable Sheep Farming,

1975 ed. Diamond Farm Book Publishers ($8.95).

Ensminger, M.E., Sheep and Wool Science, 4th ed., 1970. Interstate Publishers ($15).

Scott, George, ed., The Sheepman's Production Handbook, 1970. Sheep Industry Development Program, 200 Clayton Street, Denver, Colorado 80206 (looseleaf, $6.50). (Very useful. Can be added on to as new information is made available. For a large operation, generally, but is excellent in health, nutrition and breeding.)

Sheep Handbook, Penn. State University (paperback, $2).

Books—Breeding

Scott, George, ed. *op. cit.*

Selective Breeding for Better Sheep, Agricultural Extension Service, University of California, Davis, California 95616.

Books—Diseases

Marsh, Hadleigh, Newsom's Sheep Diseases, 1974 (reprint of 1965 edition). Robert E. Krieger Publishing Co., Inc., Box 542, Huntington, N.Y. 11743 ($15.50).

Scott, George, ed. *op. cit.*

The T.V. Vet, T.V. Vet Sheep Book, Diamond Farm Book Publishers ($9.75).

Periodicals

The Shepherd, Sheffield, Mass. 01257. $4.50/year.

Equipment

Sheepman's Supply Company, Rte. 1, Box 141, Barboursville, Virginia 22923. (Excellent catalog.)

Wool

ASCS (U.S. Department of Agriculture's Agriculture Stabilization and Conservation Service). Usually found at county seat. This agency administers the government's cost sharing program on wool.

USDA Publication: Official Standards of the United States for Wool Grades, C+M No. 135.

Wool Crop Report, New England Crop Reporting Service. USDA Statistical Report Service, Agricultural Statistician, 1305 Post Office Building, Boston, Mass. 02109. Free.

Wool—Shearing

Bowen, Godfrey, Wool Away: The Art and Technique of Shearing, 1974. Van Nostrand Reinhold (illus., paperback, $4.50).

Wool–Custom Spinning

Barlettyarns, Harmony, Maine 04942. (Send them your wool and they will send you back the same number of pounds–minus some for filth, burrs, etc.–in spun wool for about 65¢ a skein! Wide choice of colors and plys. Send for free price list and samples of colors and plys.)

Shippensburg Woolen Mills, Shippensburg, Pa. 17257. (Send your wool in and get beautiful wool blankets and other wool products in return. Write for price list and samples.)

Wool–Dyeing

Dye Plants and Dyeing. Handbook No. 46. Brooklyn Botanic Garden, 1000 Washington Avenue, Brooklyn, N.Y. 11225 ($1.50). (How to use dyes derived from plants for dyeing yarns and textiles.) Natural Plant Dyeing. Handbook No. 72. Brooklyn Botanic Garden (see above). A complement to Handbook No. 46 ($1.50).

Wool–Spinning & Weaving

Castino, Ruth, Spinning and Dyeing the Natural Way. Mother's Bookshelf ($8.95). (Spinning and dyeing using natural materials, including wild dye plants.)

Creager, Clara, Weaving: A Creative Approach for Beginners. Garden Way Publishing (paperback, $3.95).

Davenport, Elsie, Your Handspinning. Mother's Bookshelf (paperback, $4.25).

Kluger, Marilyn, The Joy of Spinning. Garden Way Publishing ($7.95).

Wool–Spinning Wheels

Mother's General Store (also knitting machines).

Velma Good, Star Rte., Box 131, Port Orford, Oregon 97465.

PIGS

Books

Johnson, Geoffrey, Profitable Pig Farming, 4th ed., Diamond Book Publishing ($9.75).

Swine Management Handbook, Penn. State University (paperback, $2).

GOATS

Books

Belanger, Jerry, Raising Milk Goats the Modern Way, 1975. Garden Way Publishing (paperback, $3.95).

Leach, Coral, Aids to Goatkeeping, 8th ed. Order from *Dairy Goat Journal*, Box 1908, Scottsdale, Arizona 85252 ($10).

Mackenzie, David, Goat Husbandry. Transatlantic Arts, Inc., North Village Green, Levittown, N.Y. 11756 ($16.75).

Walsh, Helen, Starting Right with Milk Goats, 10th printing, 1975 (originally published in 1947). Garden Way Publishing (paperback, $3.50).

Periodicals

Dairy Goat Journal, Box 1908, Scottsdale, Arizona 85252. Monthly. $5/year; $14/3 years.

Equipment

American Supply House, Box 1114, Columbia, Missouri 65201.

Breeder's Supply Company, 101 So. Main St., Council Bluffs, Iowa 51501.

Dolly Enterprises, 279 Main St., Colchester, Ill. 62326.

Hoeggen Supply Company, Box 490099, College Park, Ga. 30349 (catalog 50¢, refundable on first order).

Thiele, John H., Box 62, Warwick, N.Y. 10990.

Tomellem Company, Calico Rock, Arkansas 72519.

MISCELLANEOUS TOPICS

Breeds

Briggs, Hilton M., Modern Breeds of Livestock, 3rd. ed., 1969. Macmillan Publishing Co. (illus., $13.95).

Equipment

Jurgenson, E.M., Handbook of Livestock Equipment, 1971. Interstate Publishers ($10.75). (All about building barns, sheds, pens, feed and watering equipment, etc. Worthwhile.)

Feeding

Cassard, D.W., and Jurgenson, E.M., Approved Practices in Feeds and Feeding, 1971. Interstate Publishing (illus., $6.95).

Morrison, Frank B., Feeds and Feeding, 23rd ed., 1967. Morrison

Publishing Co., Ithaca, N.Y. 14850. (The bible on feeding. If you really want to study feeds, this is the book.)

Breeding

Ensminger, The Stockman's Handbook, 1970. Interstate Publishers. (Excellent breeding information on pigs and sheep.)

Hafez, E.S., Reproduction in Farm Animals, 3rd ed., 1974. Lea and Febiger Publishing, 600 S. Washington Square, Philadelphia, Pa. 19106 ($17.50).

Lasley, John, The Genetics of Livestock Improvement, 2nd ed., 1972. Prentice-Hall ($15.95).

Lush, Jay, Animal Breeding Plans, revised, 1962. Iowa State University Press ($10).

Health—Books

Dykstra, R.R., Animal Sanitation and Disease Control, 1961. Interstate Publishers ($14.65).

Siegmund, O.H., ed., The Merck Veterinary Manual, 4th ed., 1973. Merck & Co., Inc., Rahway, N.J. 07065 ($13.25).

Seiden, Rudolph, Livestock Health Encyclopedia, 3rd ed., 1968. Springer Publishing Co. ($13.50). (Cattle, sheep, goats, pigs, horses and mules.)

Stamm, G.W., Veterinary Guide for Farmers, 1963. Garden Way Publishing ($7.95). (Poultry, cattle, pigs and some sheep. Exhaustive but very readable.)

Health—Veterinary Supplies

As explained briefly in the appendix on giving an injection, you can save considerable amounts of money by ordering your pharmaceuticals direct from veterinary supply houses. Each of the companies listed below will furnish a free catalog. They mail orders promptly, and in an emergency some may even send off a phoned-in order the same day.

Eastern States Serum Company, 1717 Harden St., Columbia, S.C. 29204.

Kansas City Vaccine Co., Stock Yards, Kansas City, Missouri 64102.

Omaha Vaccine Co., Inc., 2900 "O" St., Omaha, Nebraska 68107.

United Pharmical Co., 8366 LaMesa Blvd., LaMesa, California 92041.

Butchering

Ashbrook, Frank G., Butchering, Processing and Preservation of Meat, 1955. Van Nostrand Reinhold Co. (paperback, $3.95). (Cattle, pigs, sheep, poultry, game and fish. Slaughter, cutting up, preserving, curing, smoking, etc. Exhaustive, but at times very hard to get through. The only all-inclusive manual, however.)

Smoking, Curing & Sausagemaking

Ashbrook, Frank G. *op. cit.*

A Complete Guide to Home Meat Curing, 1973. The Morton Salt Co. (paperback, $1). You can buy this in most feed stores.

Hertzberg, Vaughan, and Greene, Putting Food By, 2nd revised ed., 1975. The Stephen Greene Press, Brattleboro, Vt. 05301 (paperback, $4.95). (Exhaustive. A fantastic book packed with information on all aspects of "putting by," including smoking and curing, pasteurizing milk, making cottage cheese, soapmaking, etc. Buy it.)

Hull, Raymond, and Sleight, Jack, Homebook of Smoke-curing Meat, Fish and Games, 1971. Stackpole Books, Cameron and Keller Sts., Harrisburg, Pa. 17105 ($7.95).

RAK, Dept. 32, Box 4155, Las Vegas, Nevada 89106. (Free catalog of sausage recipes, curing and smoking.)

Soapmaking

Ashbrook, Frank G. *op. cit.*

Hertzberg, et al. *op. cit.*

Cheese and Butter Making

Hobson, Phyllis, Making Homemade Cheese and Butter, 1973. Garden Way Publishing (paperback, $2.50). (Making all kinds of cheese and other dairy products from cow's or goat's milk.)

Radke, Don, Cheese Making at Home: The Complete Illustrated Guide, 1974. Doubleday Publishing Co. ($5.95).

Slipper, Mitten and Clothing Patterns

For use with your tanned hides. I have had trouble finding patterns before this. It sure beats cutting up a pair of gloves to get a pattern.

The Family Creative Workshop, Vol. 10, pp. 1210-1219. Plenary Publishing International, Inc. (No need to buy the whole set; you can probably find a copy in a library.)

Farnham, A.B., The Home Manufacture of Furs and Skins, Garden Way Publishing (paperback, $2). (How to tan and, most important, how to make the skins into clothing.)

MISCELLANEOUS EQUIPMENT

Electric Fence Chargers

You can get fence chargers through any feed store or from Sears or Ward's farm catalogs. However, the best electric fencers, Gallagher's, are not available in any such place. They charge miles and miles of fencing, are strong, good weed choppers and will last forever. If you seriously want to fence large fields for sheep with electric fencing this is *the* charger for you. They are expensive, but worth it in the long run. They are made in New Zealand and the sole U.S. importers are:

Henry and Cornelia Swayze
Brookside Farm
Tunbridge, Vermont 05077
(802) 889-5581

Gristmills

Mother's General Store

Corn Shellers

Mother's General Store

Seed Broadcasters

Mother's General Store

Shredders

Winona Attrition Mill Co., Dept. CS, Box 932, Winona, Minnesota 55987. (For making silage and grinding feeds. Free literature.)

FOR FURTHER "GROW YOUR OWN–" READING:

Grains

Brickbauer, Elwood A. and Mortenson, William P., Approved Practices in Crop Production. Interstate Publishers, ($10.95). (This is a high school textbook but should offer good background.)

Delorit, Richard, et al., Crop Production, 4th ed., 1973. Prentice-Hall 07632 ($12.95).

Hughes, Harold and Metcalf, Darrell, Crop Production, 3rd ed., Illus., 1972. Macmillan Publishing Co. ($14.50).

Kipps, M.S., Production of Field Crops, 6th ed., 1970. McGraw-Hill Publishing ($18).

Lockhardt, J.A. and Wiseman, A.J., Introduction to Crop Husbandry, revised 3rd ed., 1975. Pergammon Press, Inc., Maxwell House, Fairview Park, Elmsford, N.Y. 10523 (paperback, $7).

Forages

Ahlgren, Gilbert H., Forage Crops, 2nd ed., 1956. McGraw-Hill ($13.50).

Heath, Maurice E., et al., Forages: The Science of Grassland Agriculture, 1973. Iowa State University Press, ($18.95). (The best book on forages.)

Langer, R.H., ed., Pastures and Pasture Plants, illus., 1973. Reed Books, distributed by Charles E. Tuttle, Inc., Rutland, Vt. 05701 ($19.50).

Seiden, Rudolph and Pfander, W.H., Handbook of Feedstuffs: Production, Formulation, Medication, 1957. Springer Publishing Co. ($14.95).

USDA Pamphlets (free, unless noted)
F2231 Varieties of Alfalfa
L482 Growing Crimson Clover
L532 Growing Red Clover
L484 Persian Clover: A Legume for the South
L464 Strawberry Clover: A Legume for the West
T1145 Breeding Perennial Forage Grasses
AB223 Grass Makes Its Own Food For Growth, For Forage, For Good Land Use and For Soil Conservation
AH170 Grass Varieties in the United States ($2)
F2241 Sudangrass and Sorghum-Sudangrass Hybrids for Forage
AH389 100 Native Forage Grasses in 11 Southern States ($1)
F2003 Legume Inoculation: What It Is; What It Does

Comfrey: Order from North Central Comfrey Producers, Box 195-G, Glidden, Wisconsin 54527 ($9 for 100 roots).

Collard and Kale Seeds: Order from H.G. Hastings Seed Co., Box 44088, Atlanta, Ga. 30330 or from local feed stores.

Silage:
USDA Pamphlets:
Making Grass Silage by the Wilting Method, Leaflet BDIM Inf. 38.
The Making and Feeding of Silage, Farmer's Bulletin No. 578.

Hay: Practical hay-making on a small place; order pamphlet B-15 from Garden Way Publishing Co. (50¢).

Grains and Forages:

Bromfield, Louis, Malabar Farm, 1947. Ballantine Books, N.Y., N.Y. (paperback, $1.25). (Goes into grasses a bit but is a worthwhile book to own and read in its own right. About a family and a community taking to the rural life in the forties.)

Ensminger, M.E., The Stockman's Handbook. Interstate Publishers ($16.50). (Goes into good detail on pastures, hay and silage.)

Langer, Richard, Grow It!, 1972. Avon Books, 959 8th Ave., N.Y., N.Y. 10019 (paperback, $3.95).

Rodale, J.I., et al., The Encyclopedia of Organic Gardening, 1971. Rodale Press, Emmaus, Pa. 18049 ($11.95).

Seymour, John and Sally, Farming For Self-sufficiency, 1973. Schocken Books ($7.50).

The Mother Earth News, Box 70, Hendersonville, N.C. 28739. Bi-monthly. $10/year.

Organic Gardening and Farming Magazine. (Occasional articles on the growing of field crops and forages.)

Publishers' Addresses

Diamond Farm Book
Publishing Co.
Brighton, Ontario
Canada KOK 1HO

Doubleday & Company, Inc.
245 Park Avenue
New York, N.Y. 10017

Garden Way Publishing
Company
Charlotte, Vermont 05445

The Interstate Printers &
Publishers, Inc.
19 North Jackson Street
Danville, Ill. 61832

Iowa State University Press
Ames, Iowa 50010

Macmillan, Inc.
866 Third Avenue
New York, N.Y. 10022

Pennsylvania State University
Press
215 Wagner Building
University Park, Pa. 16802

Prentice-Hall, Inc.
Englewood Cliffs, N.J. 07632

Schocken Books, Inc.
200 Madison Avenue
New York, N.Y. 10016

Springer Publishing Co.
200 Park Avenue South
New York, N.Y. 10003

Van Nostrand Reinhold Co.
300 Pike Street
Cincinnati, Ohio 45202

Glossary

BALLING GUN An instrument that is slipped partly down the throat of an animal and used to administer large pills

BARROW A male pig castrated before sexual maturity

BOAR A male pig

BOLUS GUN See BALLING GUN

BROILER A young chicken of either sex that weighs 2½ pounds and is less than eight months old (cf. FRYER and ROASTER)

BROKEN MOUTH A condition occurring in sheep and goats, usually at about the age of five or six years, whereby some of the permanent teeth are missing from the mouth (cf. FULL MOUTH, GUMMER)

BROODER The enclosure in which young animals are kept, or are allowed free access to, which furnishes heat until they adapt to outside temperatures

BROODINESS A natural condition in poultry in which a female goes out of production and sets on and attempts to hatch eggs

BROODY COOP An enclosure used to confine a broody hen and hasten her return to normal production

BUCK A male goat. A male rabbit. Less commonly: a male sheep.

CAPON A male chicken castrated surgically before sexual maturity (cf. STAG and CAPONETTE). In butchering, a castrated male chicken that weighs six to eight pounds.

CAPONETTE A male chicken that is neutered before sexual maturity by the implantation of female hormones (cf. CAPON)

CARBONACEOUS HAY See GRASS HAY

CASTRATE To remove the testicles of a male animal

CHAFF Unwanted parts of grain separated during winnowing

CHEVON Goat meat

CHICK A young chicken

CLOSED-FACED A sheep that has considerable wool covering about the face and eyes. This often leads to a condition known as wool blindness (cf. OPEN-FACED; also WOOL BLINDNESS).

COCK Any male chicken butchered after eight months of age. A stag.

COCKEREL A male chicken less than one year old

COLOSTRUM The milk produced by an animal immediately and for the first few days after the birth of its young. Rich in vitamins, minerals and antibodies. A young animal that does not receive this milk may die or will be very difficult to raise.

CREEP FEEDING The feeding of a young animal by means of an enclosure accessible to it, but not to its mother

CROP A digestive organ in poultry in which food is prepared for digestion

CROSSBRED An offspring that results from the breeding of two purebred parents of different breeds. A hybrid.

CROSSBREEDING The mating of purebred parents from different breeds (see HYBRID VIGOR)

CULL (v.) To remove an inferior animal from a flock. (n.) Any animal that is culled.

DICALCIUM PHOSPHATE A mineral mix rich in calcium and phosphorus

DISBUD To remove the horns of an animal

DOCK (v.) To cut short the tail of an animal (most commonly a lamb); usually for sanitary reasons and to facilitate breeding in females. (n.) The area around the tail of sheep or other animals.

DOE A female goat. A female rabbit.

DRAKE A male duck

DRESS OUT To remove the feathers or skin and to cut up and trim the carcass of an animal after slaughter

DRYLOT An area of confinement containing little or no natural feed into which all or most of an animal's feed must be brought (cf. FREE-RANGING)

DUCK A female duck

DUCKLING A young duck of either sex

ELASTRATOR A tool used in castrating and docking. A tight rubber band is applied to the tail or the scrotum, the circulation is thereby cut off and the tail or scrotum gradually dries up and falls off.

EMASCULATOR A tool used for docking and castrating that both cuts and crushes surrounding tissue to prevent excessive bleeding

EWE A female sheep

EYEING Clipping the wool from around the face of closed-faced sheep to prevent wool blindness

FARROW To give birth to a litter of piglets

FEED CONVERSION RATIO The rate at which an animal converts feed to meat. If an animal requires four pounds of feed to gain one pound it is said to have a four to one (4:1) feed conversion ratio.

FIELD CROPS Feed plants grown primarily for their seeds. For example, corn, wheat, oats, soybeans, etc. (cf. FORAGE CROPS).

FINISHING The increased feeding of an animal just prior to butchering, which results in rapid gains and increased carcass quality

FLAIL (n.) A tool used for threshing grain. (v.) To use a flail to thresh grain.

FLUSHING The practice of increasing the feed intake of a female animal just prior to ovulation and breeding. This causes the animal to gain some weight and drop more eggs, often resulting in larger litters.

FORAGE CROPS Those plants or parts of plants that are used for feed before maturing or developing seeds (cf. FIELD CROPS). The most common forage crop is simple pasture.

FOWL For butchering purposes, a female chicken that is more than eight months old

FREE CHOICE FEEDING A type of feeding routine whereby feed, water, salt, etc., is provided in unlimited quantities and an animal is left to regulate its intake (cf. HAND FEEDING)

FREE-RANGING Allowing animals, especially poultry, to roam freely and eat as they wish without any sort of confinement (cf. DRYLOT)

FRESHEN To come into milk, as when a dairy animal gives birth.

FRYER A chicken of either sex that weighs between 2½ and 3½ pounds and is less than eight months old (cf. BROILER and ROASTER)

FULL MOUTH A state in sheep or goats when an animal has a full set of permanent teeth. This occurs at approximately the age of four. The animal will continue to have what is known as a full mouth until it loses some teeth (see BROKEN MOUTH) or until it loses all of its teeth (see GUMMER).

GANDER A male goose

GARBAGE Scraps. Leftover food. A very poor choice of words as it has bad connotations.

GILT A female pig that has not yet produced a litter (cf. SOW)

GIRTH The measure of a distance around an animal at a point just behind the front legs

GIZZARD The muscular enlargement of the alimentary canal of poultry that immediately follows the crop. Has thick, muscular walls and a tough horny lining for the grinding of the food.

GOOSE A female goose

GOSLING A young goose of either sex

GRADE An animal of no distinguishable breed or background (cf. PUREBRED)

GRADING UP The practice of improving a flock whereby purebred sires are mated to grade animals and their offspring. In three generations the offspring will be ⅞ purebred and in some cases eligible for registration. Upgrading.

GRASS Any of the members of the family *Gramineae* (such as timothy, orchardgrass, Sudangrass, etc.). When used as pasture, hay or in silage they provide more energy than legumes but are lower in protein and vitamins (see LEGUME).

GRASS HAY Any hay totally or primarily from a grass crop (cf. LEGUME HAY; GRASS; LEGUME)

GRASS LAMB A lamb that is dropped in the springtime and is raised on pasture in the summer months and butchered in the fall when pasture dies

GRIT Any ingested rough, undigestible matter such as bits of glass or small stones that is used to grind food in the gizzard of a chicken or other poultry

GUMMER A sheep or goat having no teeth at all (cf. FULL MOUTH, BROKEN MOUTH)

HAND FEEDING A type of feeding routine whereby an animal is fed measured amounts of food, water, salt, etc. (usually just food) at fixed intervals (cf. FREE CHOICE FEEDING)

HAY Any crop (most often grasses or legumes) that is cut, dried and stored (either baled or loose) for later use

HEIFER A cow that is under three years of age and has not yet produced a calf

HEN A female chicken. A female turkey.

HETEROSIS See HYBRID VIGOR

HOTHOUSE LAMB A lamb that is dropped in the fall or early winter and is marketed at an age of 6 to 12 weeks. These are usually sold in periods before the Christmas or Easter holidays to take advantage of higher prices.

HYBRID See CROSSBRED

HYBRID VIGOR The increase of size, speed of growth and vitality of a crossbreed over its parents. Heterosis.

INBREEDING The mating of very closely related animals such as mother and son, father and daughter, brother and sister. In experienced hands it can be used to selectively maintain certain desirable traits; if used improperly it can produce undesirable traits and downgrade stock (cf. LINEBREEDING).

KID (v.) To give birth to a young goat. (n.) A young goat.

KINDLE (v.) To give birth to a litter of rabbits. (n.) A young rabbit.

LAMB (n.) A young sheep of any sex, less than one year old (cf. MUTTON, YEARLING). (v.) To give birth to a lamb (or lambs).

LAMBING LOOP A length of smooth or plastic-coated wire used as an aid in difficult lambing

LEGUME Any of the members of the family *Leguminosae* (such as clovers, alfalfa, trefoil, vetches, etc.). When used as pasture, hay or in silage, they provide more protein and vitamins than a comparable grass crop (see GRASS).

LEGUME HAY Any hay made totally or primarily from a legume crop (cf. CARBONACEOUS HAY; GRASS HAY: LEGUME)

LINEBREEDING Very similar to inbreeding but the breeding is not so close . . . for example, the mating of cousins (cf. INBREEDING)

MILK REPLACER A powder that when mixed with water is fed to young animals as the milk portion of their diet

MOLT To shed feathers, fur, skin or horns and replace them with new growth

MUTTON In butchering, any sheep over eighteen months of age (cf. LAMB, YEARLING).

NEEDLE TEETH The eight sharp teeth present in newborn piglets. Especially in large litters, these teeth can cause injury to other piglets and the sow's udder and should be clipped (see text).

OPEN-FACED Sheep that naturally have little or no wool covering about the face and eyes. This is a desirable trait as it discourages the problem of wool blindness. (See WOOL BLINDNESS; cf. CLOSED-FACED.)

PIGLET A young pig of either sex

PINNING The sticking of a young lamb's tail to its anus. This will prevent normal bowel action and result in constipation and, if not loosened in time, death.

POLLED Without horns; either naturally or by artificial means

POULT A young turkey of either sex

PULLET A young hen less than one year old

PUREBRED An animal of a recognized breed whose lineage has been kept pure (i.e., not mixed with another breed) for many generations (cf. GRADE)

PUREBREEDING The mating of purebred parents from the same breed (cf. CROSSBREEDING)

RAM An uncastrated male sheep or goat

REPLACEMENT ANIMAL A young animal that is being raised to take the place of an older animal that is being culled. E.g., since a person wants to keep a flock of five breeding ewes, he will have to buy two *replacement ewes* to take the place of the two older breeders that have died.

RINGING Clipping a breeding ram around the neck, belly and penis region in order to facilitate proper mating

ROASTER A chicken of either sex that weighs between 3½ and 5 pounds and is less than eight months old (cf. BROILER and FRYER)

ROOSTER A male chicken

RUMINANT Any one of a class of animals including sheep, goats and cows that have multiple stomachs. They are most efficient feeders because bacterial action in one of the stomachs, the rumen, increases the food value of low-grade food.

SCOURS Technically, a bacterial infection in calves and sheep that results in a whitish-yellow, foul-smelling diarrhea. Informally, any diarrhea.

SCRAPS Edible refuse that is saved from table scraps or collected from restaurants, food stores, etc., and fed to livestock.

SETTLE To become pregnant

SEX-LINKED Distinguished sexually at birth, as chicks, by differences in coloration

SHOAT See PIGLET

SILAGE A feed consisting of certain roughages and/or field

crops finely chopped, tightly packed in an airtight container and allowed to ferment in the absence of air

SOW A female pig that has produced a litter (cf. GILT)

STAG A male chicken castrated after sexual maturity (cf. CAPON). Also, any male chicken butchered after eight months of age. A cock. A male pig castrated after sexual maturity.

STRIP To remove milk from the teat of an animal by sliding the fingers from the base of the teat to the end

SWILL Scraps

TAGGING Clipping the wool from around the dock of a ewe so that it does not interfere during mating

TEASING Keeping a ram in sight of, but not in contact with, ewes just prior to breeding. This often stimulates ovulation in the ewes.

THRESH To separate grains from the plant, as in removing oats from straw

TOM A male turkey

UTERINE CAPSULE A medication administered to a ewe to prevent infection after entering her with a hand or other instrument during lambing

WATTLES Fleshy appendages hanging from the neck of a chicken, turkey or goat

WEAN To remove a young animal from its mother and accustom it to food other than its mother's milk

WETHER A castrated male lamb; a castrated male goat

WINDROW Hay or any other crop raked into a row to dry

WINNOW To remove chaff from grain by using a current of air

WOLF TEETH Needle teeth

WOOL BLINDNESS A condition that develops most often in closed-faced sheep due to irritation of the eyes by wool and particles of chaff contained therein (see CLOSED-FACED and OPEN-FACED)

YEARLING Any animal aged one year to eighteen months. (In sheep, cf. MUTTON and LAMB.)

Bibliography

Atherton, Dr. H.V. and Dodge, W.A. "Guide To Milk Sanitation Terms." Burlington: University Of Vermont Extension Service, Brieflet 1076.

Belanger, Jerome. *The Homesteader's Handbook To Raising Small Livestock*. Emmaus: Rodale Press. 1974.

"Breeds Of Swine." USDA Farmer's Bulletin No. 1263.

Bundy, Clarence E., et al. *Livestock And Poultry Production*. Englewood Cliffs: Prentice-Hall. 1975.

"Calf Plan." A Carnation-Albers publication. Los Angeles: Albers Milling Company.

Colby, Byron E. "Hints On Feeding And Managing Ewes And Lambs." Amherst: University of Massachusetts Extension Service.

______. "Suggestions On The Economical Feeding Of Sheep." Amherst: University Of Massachusetts Extension Service.

______. "Well Fed Sheep Are Profitable Sheep." Amherst: University Of Massachusetts Extension Publication.

"Culling For Higher Egg Production." Burlington: Vermont Agricultural Extension Service. Circular 115V.

"A Dairy Goat For Home Milk Production." USDA Leaflet No. 538. 1973.

"Dairy Goats: Breeding/Feeding/Management." Amherst: University Of Massachusetts. Leaflet No. 439. 1972.

Guss, Samuel B., VMD. "Don't Let Them Get Your Goat." Paper presented at 1974 meeting of Western States Veterinary Medical Association.

"Home Tanning of Sheepskins." Nova Scotia Department Of Agriculture. Made available by the Vermont Extension Service in Cooperation with the Vermont Sheepbreeders Association. 1969.

"Keeping The Newborn Lamb Alive." University of Connecticut Department of Animal Industry. Made available by the Vermont Extension Service in cooperation with the Vermont Sheepbreeders Association.

"Lambing Troubles." University of Connecticut Department of Animal Industry. Made available by the Vermont Extension Service in cooperation with the Vermont Sheepbreeders Association.

Langer, Richard. *Grow It!* New York: Avon Books. 1972.

"Livestock Waste Facilities Handbook." Iowa State University, Midwest Planning Service publication.

Mautz, William W. "Digestibility And Rate Of Passage Of Fiber In Winter Feeds By Deer." Durham: University Of New Hampshire, Animal Sciences Department. 1971.

Mercia, Leonard S. "The Family Laying Flock." Burlington: University Of Vermont Extension Service. Brieflet 1218.

———. *Raising Poultry The Modern Way.* Charlotte: Garden Way Publishing Company. 1975.

Morrison, Frank B. *Feeds And Feeding.* Morrison Publishing Company.

Nielsen, G.R. "Control Of Lice And 'Ticks' On Sheep." Burlington: USDA, University Of Vermont Cooperative Extension Publication.

"The Production And Marketing Of Sheep In New England." A New England Cooperative Extension Publication. 1970.

"Rabbits." A Carnation-Albers publication. Los Angeles: Albers Milling Company.

"Raising Ducks." USDA Farmer's Bulletin No. 2215.

"Raising Geese." USDA Farmer's Bulletin No. 2251.

"Raising Livestock On Small Farms." USDA Farmer's Bulletin No. 2224.

Robinson, Ed. "How To Raise A Pig Without Buying Feed." Charlotte: Garden Way Publishing Company. Bulletin No. B-18. 1972.

———. "Producing Eggs And Chickens With The Minimum Of Purchased Feed." Charlotte: Garden Way Publishing Company. Bulletin No. B-16. 1972.

Rodale, J.I., et al. *The Encyclopedia Of Organic Gardening.*

Emmaus: Rodale Press. 1971.
Scott, George, ed. *The Sheepman's Production Handbook.* Denver: The Sheep Industry Development Program. 1971.
Stamm, G.W. *Veterinary Guide For Farmers.* Charlotte: Garden Way Publishing Company. 1963.
"Swine." A Carnation-Albers publication. Los Angles: Albers Milling Company.
Teller, Walter M. *Starting Right With Sheep.* Charlotte: Garden Way Publishing Company. 1973.
Templeton, George. *Domestic Rabbit Production.* Danville: Interstate Printers And Publishers. 1965.
Walsh, Helen. *Starting Right With Milk Goats.* Charlotte: Garden Way Publishing Company. 1973.

Index

Your Own Notes

Your Own Notes

Your Own Notes

Your Own Notes

Your Own Notes

Your Own Notes